FURTHER

MATHEMATICS TESTS

preparing for

SELECTIVE SCHOOLS

and

SCHOLARSHIP EXAMINATIONS

* **760 multiple choice questions based on previous papers**
* **answers to all questions with outlines of solutions to harder questions**
* **ideal preparation for the Selective Schools tests in N.S.W.**
* **ideal preparation for Independent Schools entrance and scholarship examinations throughout Australia**
* **conforms to National Curriculum guidelines**

by James An, Jim Coroneos and John Smith

Contents and Score

FURTHER MATHEMATICS TESTS

PAPER 1

1	$\frac{12}{18} = \frac{\square}{3}$. The missing number in the box is	(A) 1 (B) 2 (C) 5 (D) 6
2	Divide 8 by $\frac{1}{3}$. The answer is	(A) 24 (B) $\frac{3}{8}$ (C) $2\frac{2}{3}$ (D) $\frac{1}{24}$
3	6.4 - 2.91 =	(A) 9.31 (B) 3.51 (C) 3.49 (D) 4.49
4	$(0.3)^2 =$	(A) 0.6 (B) 0.9 (C) 0.09 (D) 0.06
5	A student finds the sum of the squares of 6 and 8. He then finds the square root of the answer. The answer should now be	(A) 14 (B) 10 (C) 100 (D) None of these
6	How many tenths are there in one half?	(A) 4 (B) 5 (C) 10 (D) 1
7	$2\frac{3}{4}$ hours = ... mins.	(A) 180 (B) 165 (C) 170 (D) 140
8	$5\frac{2}{5}$ m is the same as	(A) 5m 40cm (B) 5m 20cm (C) 5m 50cm (D) 5m 5cm
9	The diagram shows the net of a tetrahedron. What fraction of the net is shaded? 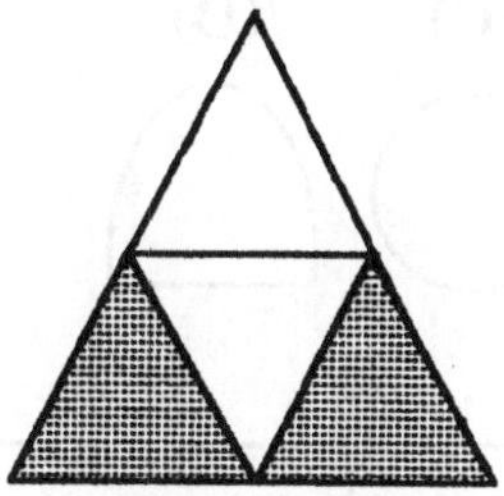	(A) $\frac{2}{3}$ (B) $\frac{1}{3}$ (C) $\frac{1}{4}$ (D) $\frac{1}{2}$
10	19% means 19 out of	(A) 10 (B) 19 (C) 90 (D) 100

11	Express 0.44 in percentage form.	(A) 0.44% (B) 4.4% (C) 44% (D) 440%

12 Which of the following could be a trapezium only?

(A) (B) (C) (D)

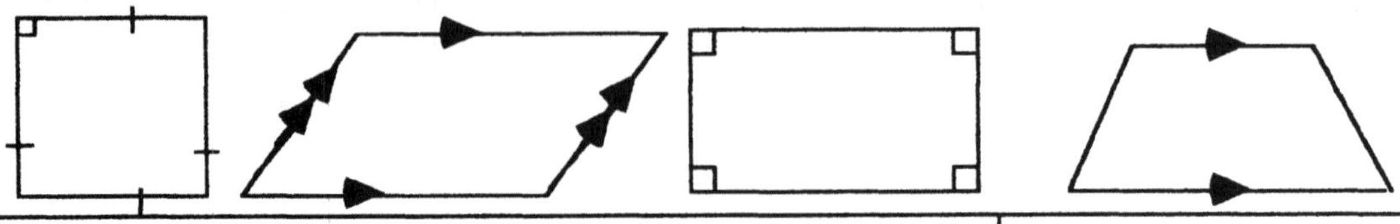

13	A helicopter flying north turned clockwise through 225°. In what direction is it now flying?	(A) SE (B) SW (C) NE (D) NW

14 - 15

The children in a school voted on their favourite foods. The results are shown on this sector graph.

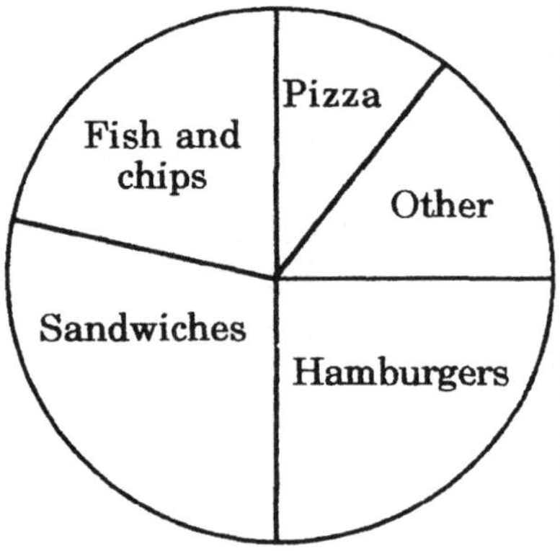

Fish and chips
Pizza
Other
Sandwiches
Hamburgers

14	Which food was most popular?	(A) Fish and chips (B) Hamburgers (C) Pizza (D) Sandwiches
15	About what percentage of children preferred hamburgers?	(A) 20% (B) 25% (C) 50% (D) 75%

16 Match the cross-sectional shape with the vertical cut indicated on the cone.

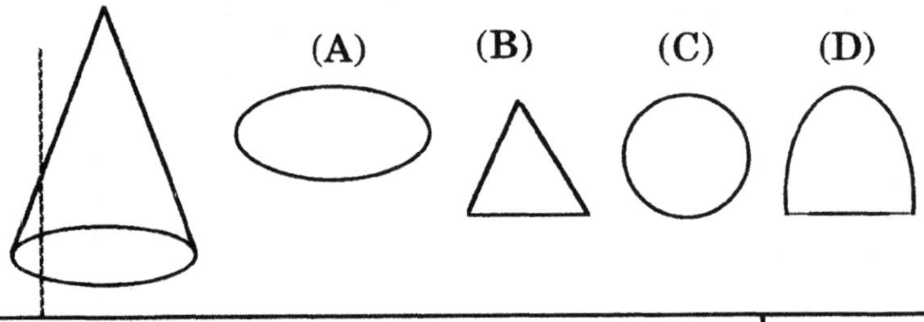

(A) (B) (C) (D)

17	Which of the following solids could have a top view of ☒ ?	(A) Cone (B) Triangular prism (C) Square pyramid (D) Cube

18	The number 7529 can be written as	(A) 750 + 29 (B) 700 + 50 + 20 + 9 (C) 7000 + 52 + 9 (D) 7000 + 500 + 20 + 9
19	The digit 6 in the number 4 176 230 represents	(A) 600 (B) 60 (C) 6 (D) 6000
20	How many kilograms in 0.65 tonnes?	(A) 65 (B) 650 (C) 1538 (D) 350
21	In a trip lasting 4hr 20min, a boat averaged 15 knots. How many nautical miles was the trip? [Note: 1 knot = 1 nautical mile per hour.]	(A) $19\frac{1}{3}$ (B) 63 (C) 65 (D) $17\frac{1}{3}$
22	In a parallelogram, the opposite angles	(A) are unequal (B) are acute (C) are equal (D) are right angles
23	$\frac{3}{5}$ of 4 km = ... m	(A) 24 (B) 240 (C) 2400 (D) 24 000
24	If 1cm^3 of water weighs 1 gram, what is the weight of 1 litre of water?	(A) 1000kg (B) 100kg (C) 10kg (D) 1kg
25	A train leaves a station at 9.15p.m. and arrives at its destination at 2.35p.m. the next day. How long does it take for the journey?	(A) 17h 20min (B) 5h 20min (C) 6h 40min (D) 11h 50min
26	The digits of the number 4179 are reversed to form a new number. The smaller 4-digit number is then subtracted from the larger. The answer is	(A) 5535 (B) 5355 (C) 9714 (D) None of these
27	Seven more than twice the sum of 6 and 3 is	(A) 25 (B) 81 (C) 22 (D) 47

28	When the product of the digits of the number 295 is added to the sum of the digits the answer is	(A) 97 (B) 106 (C) 93 (D) 115
29	Julie has enough money to buy 2kg of cherries at \$7.50 per kg. With this money she can buy 25 apples instead of the cherries. Find the cost of each apple.	(A) 50c (B) 60c (C) 70c (D) \$1
30	For the solid shown, the letter V stands for the number of vertices, F for the number of faces and E for the number of edges. The value of V + F + E is	(A) 26 (B) 28 (C) 32 (D) 34
31	As a fraction of 1 week, 63 hours is	(A) $\frac{1}{3}$ (B) $\frac{1}{4}$ (C) $\frac{3}{8}$ (D) $\frac{5}{8}$
32	A girl walks 50m in a minute. At the same rate, how long would it take her to walk a kilometre?	(A) 20 min (B) 10 min (C) 30 min (D) 25 min
33 - 34	A rectangle is 400m long and 300m wide.	
33	The perimeter in km is	(A) 1400 (B) 1.4 (C) 14 (D) 0.7
34	The area in hectares is	(A) 1.2 (B) 12 (C) 120 (D) 0.12
35	The average mass of four parcels is 2.5 kg. If three of the parcels are of mass 2.9kg, 1.3kg, 3.7kg respectively, what is the fourth mass?	(A) 3.1kg (B) 2.6kg (C) 2.1kg (D) 7.9kg
36	The volume of a closed cube is $27cm^3$. What is the total surface area in cm^2?	(A) 18 (B) 81 (C) 486 (D) 54

37	A flat is to be paid off by 100 monthly instalments of $1200. How much will still have to be paid after 4 years?	(A) $57 600 (B) $177 600 (C) $62 400 (D) $120 000
38	Instead of multiplying by 2, Lee divided by 3. Lee's answer was 48. The correct answer should have been	(A) 32 (B) 96 (C) 72 (D) 288
39	Four students were talking about averages. Only one of their statements was correct. Which one is it? (A) Student 1: The average of 3 numbers is halfway between the smallest and largest numbers. (B) Student 2: If you multiply the average of 6 numbers by 6, you get the sum of these numbers. (C) Student 3 : The average is never the same as any of the numbers you are averaging. (D) Student 4: You get the average by halving the difference between the largest and smallest numbers.	
40	After William has given $10 to Alexander, they have equal sums of money. What is the difference between their original sums of money?	(A) $10 (B) $20 (C) $30 (D) $40

PAPER 2

1	Which one of the following numbers is nearest to 5?	(A) 4.97 (B) 4.79 (C) 5.02 (D) 5.21
2	What is the difference between the value of the first 3 and that of the second 3 in the number 3.23?	(A) 100 (B) 29.7 (C) 2.97 (D) 10
3	Which one of the following is nearest to 4.25 x 4 ?	(A) 4 x 4 (B) 5 x 3.5 (C) 6 x 3.25 (D) 7 x 2.4
4	38% means .... out of 50	(A) 76 (B) 38 (C) 24 (D) 19
5	$\frac{10}{10}$ expressed as a percentage is	(A) 110% (B) 100% (C) 10% (D) 1%
6	A triangle with 2 equal sides only is called ....triangle.	(A) a scalene (B) an isosceles (C) a right-angled (D) an equilateral

7 - 9 Look at the map carefully and answer questions below.

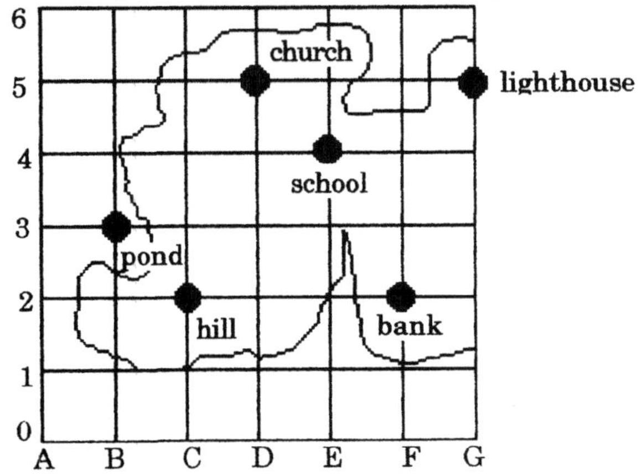

7	What is found at E4?	(A) bank (B) school (C) hill (D) pond
8	Give the coordinates that represent the church?	(A) D4 (B) E4 (C) D5 (D) E5

9 Which of the following is the nearest to the school?

(A) hill
(B) bank
(C) lighthouse
(D) church

10 - 11 The following table shows the maximum and minimum temperatures during a week in Darwin. Study it and answer the following questions.

	Minimum (°C)	Maximum (°C)
Sunday	22	33
Monday	20	31
Tuesday	21	33
Wednesday	19	34
Thursday	24	36
Friday	23	33
Saturday	19	30

10 Which day was hottest?

(A) Sunday
(B) Tuesday
(C) Thursday
(D) Friday

11 On which day was the change in temperature least?

(A) Thursday
(B) Friday
(C) Saturday
(D) Sunday

12 There are 240 pages in a book. John has finished reading $\frac{2}{3}$ of them. How many more pages does he have to read to complete the book?

(A) 80
(B) 180
(C) 120
(D) 160

13 A girl can save $60 in one year. How much can she save in 5 months, assuming she saves the same amount each month?

(A) $300 (B) $30
(C) $50 (D) $25

14 Three similar chairs costs $19.50. How many such chairs can be bought with $130?

(A) 20 (B) 40
(C) 50 (D) 58

15 Mr. and Mrs. Wood and their two children went on a holiday. Mr. and Mrs. Wood spent $980 each and each child spent $460. How much did they spend altogether?

(A) $2420
(B) $2880
(C) $1440
(D) $1900

16 Which of the following is the smallest?

(A) 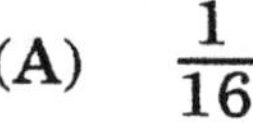 $\frac{1}{16}$ (B) 16%
(C) 60% (D) 0.075

17 Which one of the angles below is more than 2 right angles?

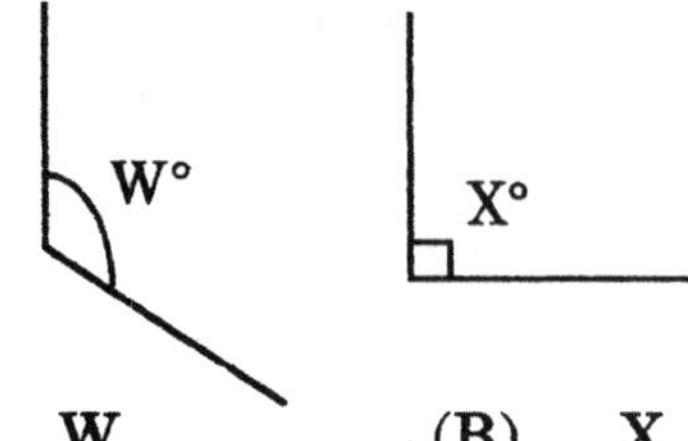

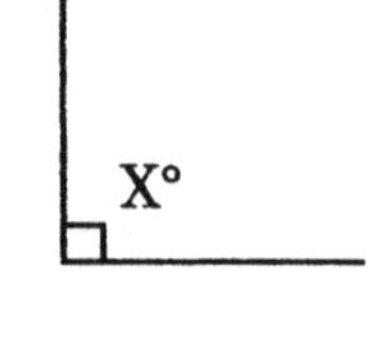

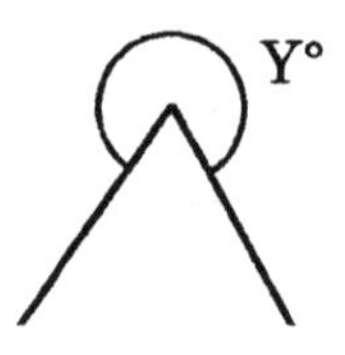

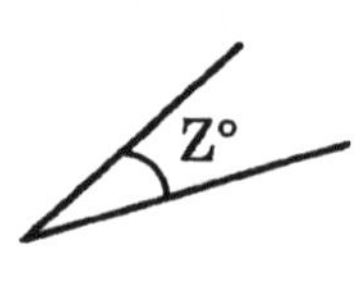

(A) W (B) X (C) Y (D) Z

18 Name the solid that can be formed from the net shown.

(A) Triangular prism
(B) Triangular pyramid
(C) Triangle
(D) None of these

19 Which of the following diagrams does not represent the net of a cube?

(A)
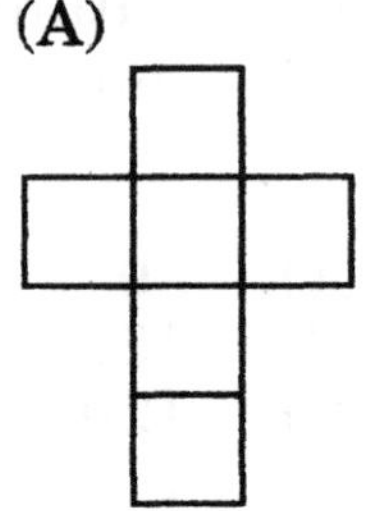
(B)
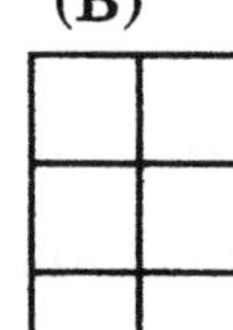
(C)
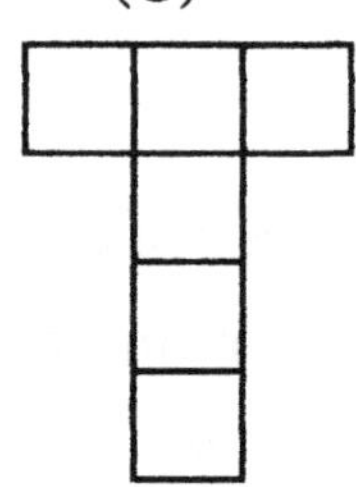
(D)
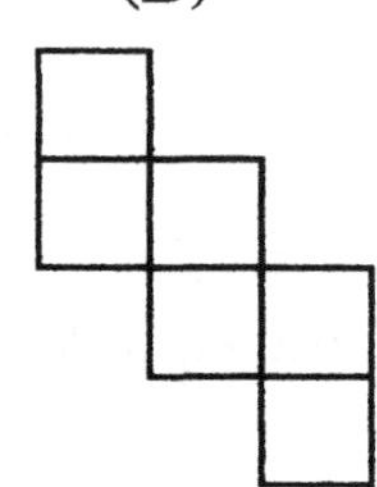

20 If I doubled the product of 7 and 3 and subtracted double the sum of 7 and 3, I would get

(A) 0 (B) 22
(C) 25 (D) 29

21 When 4 is added to the square of a certain number, the result is 53. The number could have been

(A) 7
(B) 49
(C) 6
(D) None of these

22 The smallest number which is divisible by 36 and 48 is

(A) 72 (B) 96
(C) 144 (D) 288

23 Instead of adding 7, a student subtracted 7. The student's answer was 17. The correct answer was	(A) 31 (B) 3 (C) 24 (D) 27
24 Jenny has a V.C.R. which has a 24 hour clock. She wishes to record a T.V. programme lasting 140 minutes. The programme begins at 7.35 p.m. What time will the V.C.R. show when the programme is finished?	(A) 21:25 (B) 21:55 (C) 20:75 (D) 09:55
25 A train covers 24 km in 20 minutes. How far will it go in $2\frac{1}{4}$ hours?	(A) 162 km (B) 180 km (C) 144 km (D) 216 km
26 What decimal of a metre is 15 mm?	(A) 0.015 (B) 0.15 (C) 0.0015 (D) 1.5
27 Sam is 105cm tall. His brother Ben is $\frac{2}{7}$ as tall again as Sam. How tall is Ben?	(A) 30cm (B) 135cm (C) 120cm (D) 75cm
28 How many 375 mL bottles of juice can be filled from a vat containing 30 litres of the juice?	(A) 80 (B) 160 (C) 120 (D) 11 250
29 How many grams are there in 2.7 tonnes?	(A) 27 000 (B) 270 000 (C) 2 700 000 (D) 27 000 000
30 Hugo is a wrestler. If $\frac{3}{8}$ of his mass is 48kg, what is his mass?	(A) 18kg (B) 128kg (C) 144kg (D) 384kg
31 - 32 A rectangle is 18m long and 8m wide. 31 What is the length of the side of the square whose perimeter is the same as that of the rectangle?	(A) 6.5m (B) 13m (C) 26m (D) 52m
32 What is the length of the side of the square whose area is the same as that of the rectangle?	(A) 144m (B) 72m (C) 12m (D) 13m

33 A rectangular box has length 8m and breadth 6m. If its volume is $24m^3$, what is its height?	(A) 2m (B) 1m (C) 50cm (D) 25cm
34 - 35 In the sketch showing a number of cubes each of side 1cm, the cube with the face shaded and the ones directly below it are removed.	
34 The surface area of the remaining solid, in cm^2, is	(A) 54 (B) 52 (C) 64 (D) 58
35 The volume of the remaining solid, in cm^3, is	(A) 26 (B) 27 (C) 24 (D) 25
36 What shape am I? I have 5 faces, 4 of which are triangles and the other one is a square.	(A) triangular pyramid (B) triangular prism (C) square prism (D) square pyramid
37 - 39 The travel graph shows a cyclist and a marathon runner who are on the same road between towns P and Q. The cyclist leaves town P at 11a.m. to go to town Q.	
37 The average speed of the cyclist for the journey is	(A) 18km/h (B) 20km/h (C) 22.5km/h (D) 25km/h

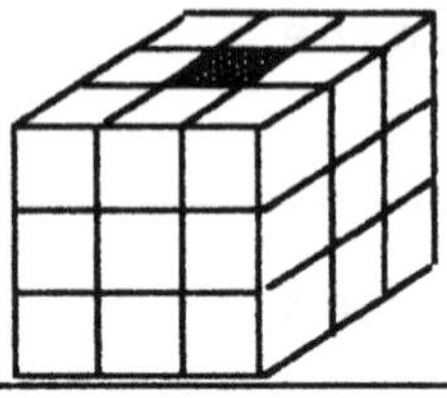

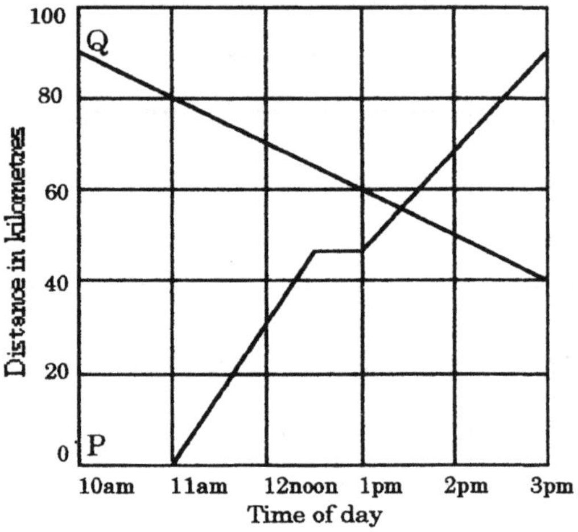

38 The approximate time and distance from Q when they pass one another is

(A) 1:30p.m., 55km
(B) 1:15p.m., 45km
(C) 1:45p.m., 50km
(D) 1:30p.m., 35km

39 If the runner stops for 30 minutes at 3 p.m., and then completes the run from Q to P at the same speed as before, at what time will the runner reach P?

(A) 7p.m.
(B) 6:30p.m.
(C) 8:30p.m.
(D) 7:30p.m.

40 A student is working on the following table:

Row	Counting Numbers	Sum of counting numbers	Product of counting numbers	Product is divisible by sum
1	1	1	1	Yes
2	1, 2	3	2	No
3	1, 2, 3	6	6	Yes

The next two rows for which the answer is "Yes" will be

(A) 4 and 5
(B) 5 and 7
(C) 5 and 6
(D) None of these

FURTHER MATHEMATICS TESTS

PAPER 3

1	What fraction is 4kg of 1 tonne?	(A) $\frac{1}{25}$ (B) $\frac{1}{250}$ (C) $\frac{2}{5}$ (D) $\frac{1}{2500}$
2	0.2 x 0.3 x 0.4 =	(A) 2.4 (B) 0.24 (C) 0.024 (D) 0.9
3	$\sqrt{9+16} =$	(A) 7 (B) 5 (C) 25 (D) 625
4	Simplify $\frac{1}{3} \times \frac{7}{8}$	(A) $\frac{8}{11}$ (B) $\frac{29}{24}$ (C) $\frac{21}{11}$ (D) $\frac{7}{24}$
5	The number for 4 hundreds 6 tenths 2 hundredths is	(A) 4.62 (B) 460.02 (C) 400.62 (D) 462
6	If you are facing south, what direction is to your right?	(A) North (B) South (C) East (D) West
7	$1\frac{3}{4} - \square = \frac{3}{8}$. The missing number in the box is	(A) $1\frac{1}{8}$ (B) $1\frac{3}{8}$ (C) $1\frac{1}{2}$ (D) $2\frac{1}{8}$
8	Add $\frac{1}{2}$ to $\frac{1}{4}$. The answer expressed as a percentage is	(A) 750% (B) 75% (C) 7.5% (D) 0.75%
9	A cyclist travels 57km per day. How many days will he take to go 1026km?	(A) 18 (B) 28 (C) 24 (D) 26
10	3km 750m is $\frac{5}{8}$ of	(A) 10km (B) 3km (C) 60km (D) 6km

11	The net shown folds to form a cylinder. The circumference of each circular end is 15cm and the length is 20cm. What is the size of PQ? 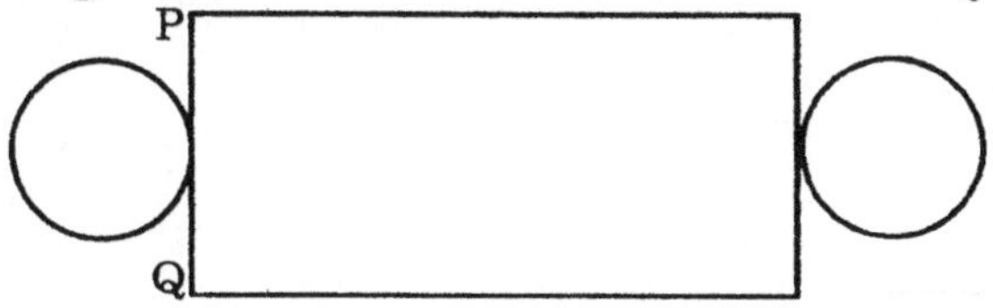	(A) 15cm (B) 20cm (C) 35cm (D) $17\frac{1}{2}$cm
12	Each month a woman saves $\frac{2}{7}$ of her income and spends \$800. What is her yearly income?	(A) \$13 440 (B) \$1120 (C) \$5600 (D) \$6857
13	5 kg of flour cost \$12.50 and 3kg of sugar cost \$3.60. Find the total cost of 3kg of flour and 4kg of sugar.	(A) \$3.70 (B) \$8.70 (C) \$12.30 (D) \$16.10
14	A sum of money is divided equally among a group of 24 people, and each gets \$25. How much will each person get if six more people join the group and share the same amount of money?	(A) \$15 (B) \$20 (C) \$25 (D) \$30
15	A book costs \$3. One free copy is given to a customer as a gift for every 4 of these books bought. If Sandra spends \$51 on buying these books, how many books will she get altogether?	(A) 24 (B) 22 (C) 21 (D) 20
16	Subtract 149 from the product of 12 and 12.5. The answer expressed as a percentage of 20 is	(A) 5% (B) 20% (C) 1% (D) 25%
17	Which of the following three angles could be the angles of a triangle?	(A) 40°, 70°, 80° (B) 45°, 60°, 45° (C) 25°, 75°, 90° (D) 40°, 90°, 50°
18	In the figure, the value of angle X + angle Y = 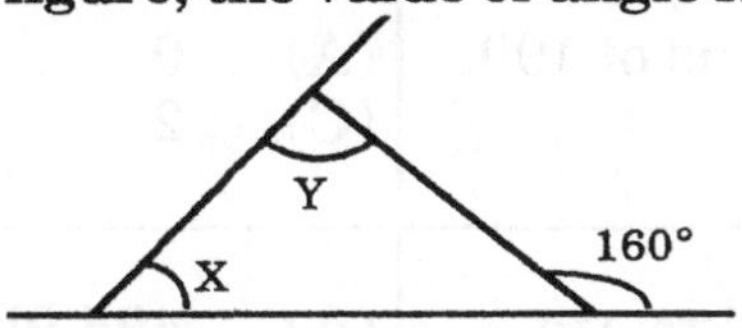	(A) 160° (B) 90° (C) 20° (D) 110°

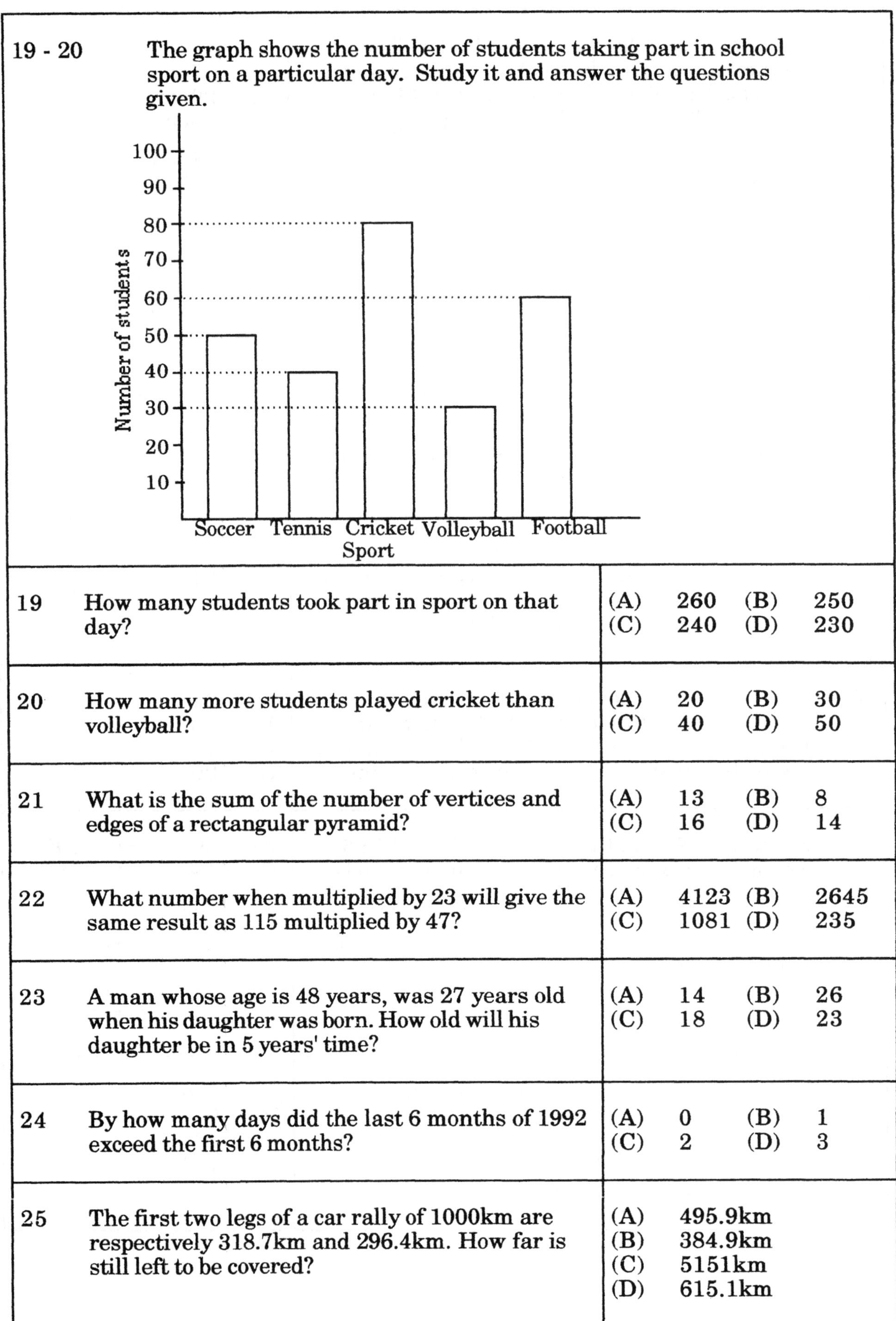

19 - 20 The graph shows the number of students taking part in school sport on a particular day. Study it and answer the questions given.

19	How many students took part in sport on that day?	(A) 260 (B) 250 (C) 240 (D) 230
20	How many more students played cricket than volleyball?	(A) 20 (B) 30 (C) 40 (D) 50
21	What is the sum of the number of vertices and edges of a rectangular pyramid?	(A) 13 (B) 8 (C) 16 (D) 14
22	What number when multiplied by 23 will give the same result as 115 multiplied by 47?	(A) 4123 (B) 2645 (C) 1081 (D) 235
23	A man whose age is 48 years, was 27 years old when his daughter was born. How old will his daughter be in 5 years' time?	(A) 14 (B) 26 (C) 18 (D) 23
24	By how many days did the last 6 months of 1992 exceed the first 6 months?	(A) 0 (B) 1 (C) 2 (D) 3
25	The first two legs of a car rally of 1000km are respectively 318.7km and 296.4km. How far is still left to be covered?	(A) 495.9km (B) 384.9km (C) 5151km (D) 615.1km

26 6.41kg = ❑ mg. What is the value of ❑?	(A) 6410 (B) 64 100 (C) 641 000 (D) 6 410 000
27 If 1kg costs \$18, what is the cost of 650g?	(A) \$10.80 (B) \$9.75 (C) \$27.69 (D) \$11.70
28 A girl's pace is 80cm. How many paces does she take in stepping out a distance of 1km?	(A) $1\frac{1}{4}$ (B) $12\frac{1}{2}$ (C) 1250 (D) 12 500
29 Pedr walked $13\frac{3}{4}$ km in 2 hr 30min. What was his average speed?	(A) $6\frac{7}{8}$ km/h (B) $34\frac{3}{8}$ km/h (C) $5\frac{1}{2}$ km/h (D) $4\frac{1}{2}$ km/h
30 - 31 The area of a square is 144cm^2. 30 What is the perimeter of the square?	(A) 24cm (B) 72cm (C) 36cm (D) 48cm
31 What is the area of the rectangle whose length is that of the side of the square increased by 3cm and whose breadth is that of the side of the square diminished by 5cm?	(A) 63cm^2 (B) 105cm^2 (C) 153cm^2 (D) 255cm^2
32 All rows, columns and diagonals in the given Magic Square add to the same number. [8, ,] [, 5, 7] [, 9,] By completing the Magic Square, you can see that this number must be	(A) 16 (B) 17 (C) 13 (D) 15
33 The value of $3^2 + 2^3$ is	(A) 17 (B) 3125 (C) 12 (D) 72
34 Fred, Ivan and Kim play cricket. Together they score 222 runs. Fred and Kim together score 193 runs and Ivan and Kim score 159 runs between them. How many runs does Kim score?	(A) 30 (B) 29 (C) 63 (D) None of these

35	I have two numbers. I form their product and their sum and find that the difference between their product and their sum equals the difference between the two original numbers. The two numbers could be	(A) 9 and 4 (B) 6 and 3 (C) 7 and 2 (D) 5 and 4
36	A bag contains 250 marbles. Three fifths of the marbles are divided into 3 equal groups. If a marble weighs 2g, find the weight of the marbles in each group.	(A) 50g (B) 100g (C) 150g (D) 250g
37	A student is completing the following table: Row A: 1, 2, 3, 4, 5, ... Row B: 1, 2, 4, 8, 16, ... Row C: 2, 4, 7, 12, 21, ... The first entry in Row C which is greater than 100 is most likely to be	(A) 136 (B) 135 (C) 137 (D) 265
38	One tap could fill a bucket in 3 minutes. Another tap could fill the same bucket in 2 minutes. How long would it take to fill the bucket if both taps were used?	(A) 1 minute 12 seconds (B) 50 seconds (C) 5 minutes (D) 6 minutes
39	A certain number is divided by 3. The answer is then multiplied by 6 and the answer to this is then halved. If you then add 5 and subtract the original number, you will get	(A) 23 (B) 5 (C) 13 (D) there is not enough information to tell
40	The product of three consecutive numbers is 720. The sum of the three numbers must be	(A) 17 (B) 9 (C) 21 (D) None of these

PAPER 4

1	The value of $\frac{3}{4} \div 9$ is the same as	(A) $\frac{4}{3} \times \frac{1}{9}$ (B) $\frac{3 \div 9}{4 \div 9}$ (C) $\frac{4}{3} \times 9$ (D) $\frac{3}{4} \times \frac{1}{9}$
2	40 - 8 x 2 + 3 =	(A) 21 (B) 0 (C) 27 (D) 67
3	A tank holds 360 litres of oil. How much oil is in the tank when it is three quarters full?	(A) 90L (B) 45L (C) 135L (D) 270L
4	50% of 3m is	(A) 150m (B) 15m (C) 1.5m (D) 50cm
5	20% is the same as 10 out of	(A) 10 (B) 20 (C) 40 (D) 50
6	How many persons can be paid $47 each from a fund of $908?	(A) 20 (B) 19 (C) 38 (D) 24
7	The product of 4 and the sum of 2 and 5 is	(A) 13 (B) 22 (C) 24 (D) 28
8	3.7kg is equivalent to	(A) 3700t (B) 370 000mg (C) 370g (D) 0.0037t
9	The area of a rectangle is $3400m^2$. What is its area in ha?	(A) 3.4 (B) 34 (C) 0.34 (D) 0.034
10	How many multiples of 5 are there which are less than 70, greater than 0 and divisible by 3?	(A) 14 (B) 13 (C) 23 (D) 4

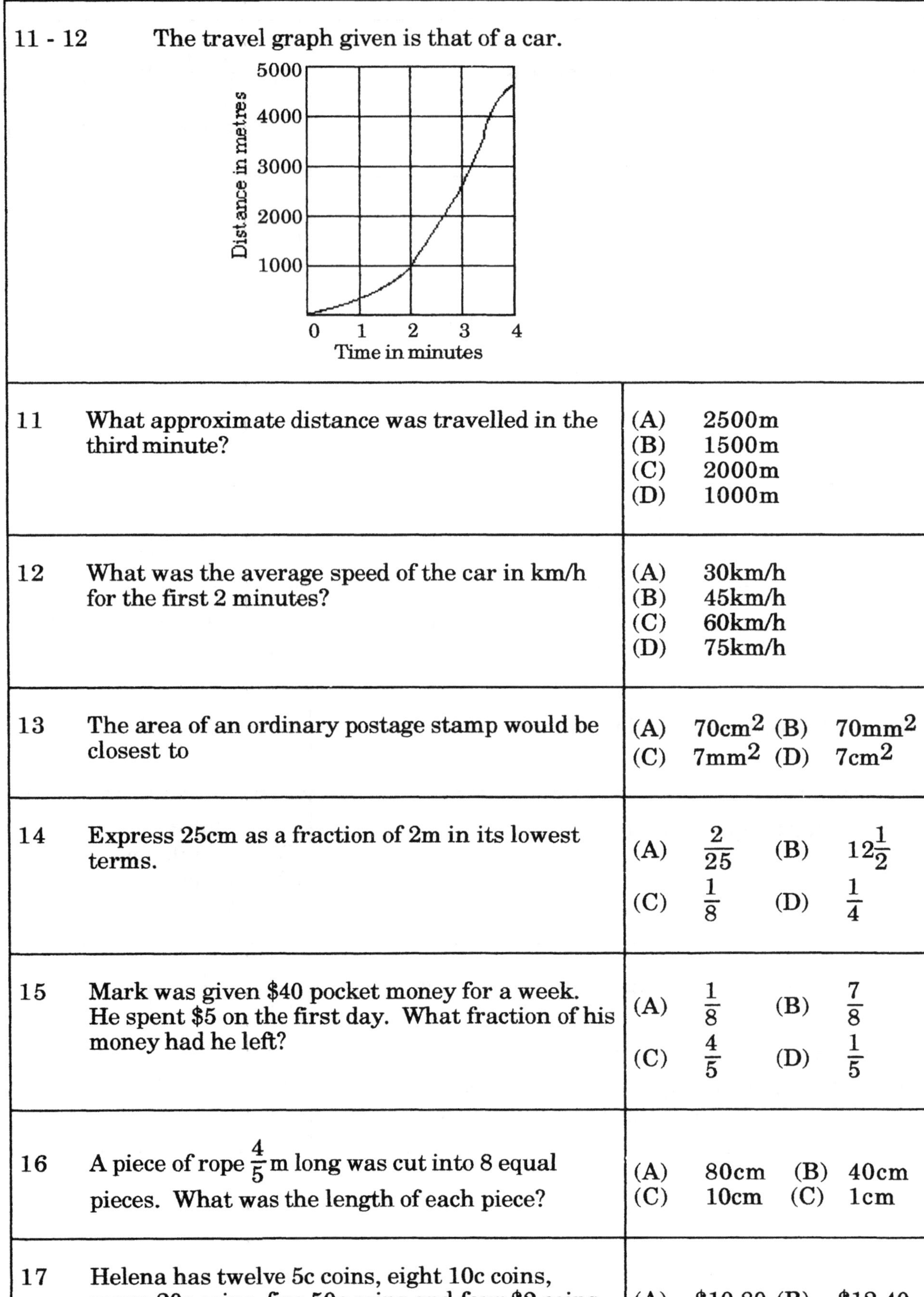

11 - 12 The travel graph given is that of a car.

11	What approximate distance was travelled in the third minute?	(A) 2500m (B) 1500m (C) 2000m (D) 1000m
12	What was the average speed of the car in km/h for the first 2 minutes?	(A) 30km/h (B) 45km/h (C) 60km/h (D) 75km/h
13	The area of an ordinary postage stamp would be closest to	(A) 70cm^2 (B) 70mm^2 (C) 7mm^2 (D) 7cm^2
14	Express 25cm as a fraction of 2m in its lowest terms.	(A) $\frac{2}{25}$ (B) $12\frac{1}{2}$ (C) $\frac{1}{8}$ (D) $\frac{1}{4}$
15	Mark was given \$40 pocket money for a week. He spent \$5 on the first day. What fraction of his money had he left?	(A) $\frac{1}{8}$ (B) $\frac{7}{8}$ (C) $\frac{4}{5}$ (D) $\frac{1}{5}$
16	A piece of rope $\frac{4}{5}$m long was cut into 8 equal pieces. What was the length of each piece?	(A) 80cm (B) 40cm (C) 10cm (C) 1cm
17	Helena has twelve 5c coins, eight 10c coins, seven 20c coins, five 50c coins and four \$2 coins. Helena's coins add up to	(A) \$10.80 (B) \$12.40 (C) \$12.50 (D) \$13.30

18	The daily charge for parking a car in a building is \$5 per hour (or part of an hour). If Mr. Chan parks his car from 11:00a.m. to 4:30p.m. on one day and 12:30p.m. to 3:00p.m. on another, how much will he have to pay altogether?	(A) \$40 (B) \$45 (C) \$54 (D) \$60
19	Father bought a belt for \$36.90 and two pairs of socks at \$7.50 each. What amount of change would he receive if he paid with three \$20 notes?	(A) \$7.50 (B) \$8.10 (C) \$9.40 (D) \$10.00
20	Express 0.65kg as a percentage of 1.25kg.	(A) 65% (B) 52% (C) 50% (D) 48%
21	Which one of the following is the condition for two lines to be parallel? (A) They intersect at a distant point. (B) They are of the same length. (C) They are drawn from the same point. (D) The perpendicular distance between them is always the same.	
22	If AB is parallel to CD and CD is parallel to XY, then (A) AB is equal to XY (B) AB is perpendicular to XY (C) AB is parallel to XY (D) CD is longer than both AB and XY	
23	A man leaves his house. He runs 4km due north, then east for 8km, then south for 4km. How far is he from his house?	(A) 4km (B) 8km (C) 12km(D) 6km
	The graph shows the distances covered by four boys A, B, C, and D during a bicycle race. A B C D Distance Time	
24	Which one of them is the fastest cyclist?	(A) A (B) B (C) C (D) D

25 One thousand seconds is approximately	(A) 1 hour (B) 20 minutes (C) 10 minutes (D) 60 hours
26 A group of people at a restaurant receive a bill for \$377. They decide to split the bill evenly. One man pays a total of \$174 for himself and 5 others. How many people were in the group?	(A) 13 (B) 19 (C) 29 (D) None of these
27 - 29 A method of measuring weight, called Troy Weight, is given below: 24 grains make 1 pennyweight 20 pennyweights make 1 ounce 12 ounces make 1 pound. Use this table to answer the questions below:	
27 The number of grains in 12 pennyweights is	(A) 2 (B) $\frac{1}{2}$ (C) 144 (D) 288
28 The number of ounces in 2880 grains is	(A) 24 (B) 48 (C) 6 (D) 8
29 The number of grains in $\frac{1}{4}$ of a pound is	(A) 160 (B) 48 (C) 1440 (D) 720
30 How long does a car take to cover 60km at 75km/h?	(A) 48min (B) 75min (C) 40min (D) 54min
31 How many cm in 0.7 of 2.4m?	(A) 1680 (B) 168 (C) 148 (D) 16.8
32 - 33 A rectangle P has breadth 12cm and area 192cm^2, whilst a square Q has the same perimeter as P. 32 What is the perimeter of P?	(A) 28cm (B) 32cm (C) 56cm (D) 64cm
33 The area of P is (A) is the same as that of Q (B) exceeds that of Q by 4cm^2 (C) is less than that of Q by 4cm^2 (D) cannot be determined	

34	The mass of Kim is 45.6kg and of her dog Fido is 2.4kg. What would be their total mass on the Moon if an object weighs 6 times as much on Earth as it does on the Moon?	(A) 8kg (B) 78.4kg (C) 48kg (D) 288kg
35	Two glasses together contain 280mL of water. The first glass contains 80mL more than the second glass. How much water does the first glass contain?	(A) 200mL (B) 140mL (C) 100mL (D) 180mL
36	A hockey field is 100m from goal line to goal line. If it is 320m around the field, what is its width?	(A) 120m (B) 60m (C) 220m (D) 110m
37 - 39	The sketches show three nets each consisting of 3cm squares. 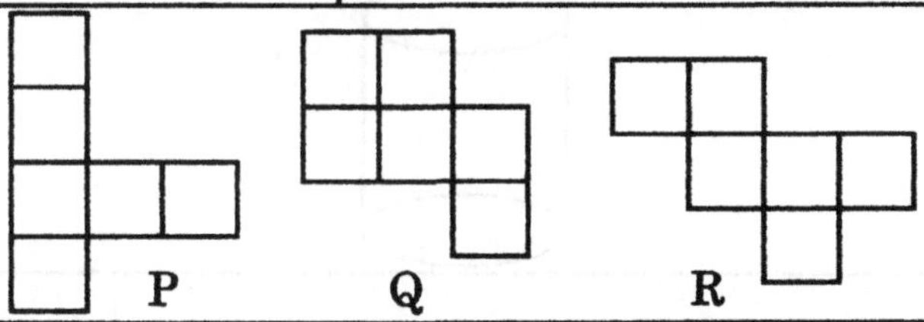 (Nets P, Q, R)	
37	Which of these nets cannot be folded to form a cube?	(A) P and Q only (B) Q and R only (C) P and R only (D) Q only
38	The total internal and external surface area of any cube which can be constructed from any of the above nets is	(A) $54cm^2$ (B) $108cm^2$ (C) $72cm^2$ (D) $144cm^2$
39	The volume of this cube is	(A) $27cm^3$ (B) 9cm (C) $54cm^3$ (D) $81cm^3$
40	In the solid drawn, if V stands for the number of vertices, F the number of faces, and E for the number of edges, then the value of V + E - F is 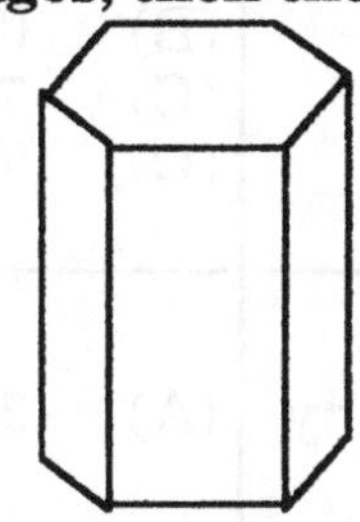	(A) 22 (B) 2 (C) 18 (D) 20

FURTHER MATHEMATICS TESTS

PAPER 5

1	68% expressed as a decimal number is	(A) 0.68 (B) 6.8 (C) 68 (D) 680
2	How many axes of symmetry has a regular pentagon?	(A) 1 (B) 3 (C) 5 (D) 7
3	Which solids have been joined to form the solid pictured? 	(A) cone and sphere (B) cylinder and circle (C) sphere and cylinder (D) cylinder and cone
4	The sum of five thousand and forty one and three thousand two hundred and five is	(A) 8246 (B) 8606 (C) 8066 (D) 8426
5	The difference in the value of the two 7' s in the number 7307 is	(A) 0 (B) 693 (C) 723 (D) 6993
6	What time will it be when I reach school if I leave home at 26 minutes past 7 in the morning and the journey takes me 45 minutes?	(A) 8:01 a.m. (B) 7:56 a.m. (C) 8:11 a.m. (D) 8:21 a.m.
7	What decimal of 1 t is 23kg?	(A) 0.23 (B) 0.023 (C) 2.3 (D) 0.0023
8	Only one of the following equals 8. Which is it?	(A) $7 \times 7 + 7 \div 7$ (B) $(7 \times 7 + 7) \div 7$ (C) $7 \times (7 + 7) \div 7$ (D) $7 \times (7 + 7 \div 7)$
9	A container can hold $20\frac{1}{8}$ L of water. If it already contains $6\frac{3}{8}$ L of water, how much more water can be poured into it to fill it completely?	(A) $26\frac{1}{2}$ L (B) $14\frac{1}{2}$ L (C) $13\frac{1}{4}$ L (D) $13\frac{3}{4}$ L

10 Mrs Wilson had \$400. She spent $\frac{5}{8}$ of her money in a shop. How much money had she left?

(A) \$250 (B) \$150
(C) \$100 (D) \$50

11 Which rectangle has its shaded area equal to $\frac{1}{3}$ of its total area?

(A) 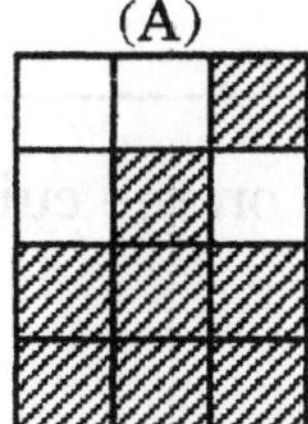
(B) 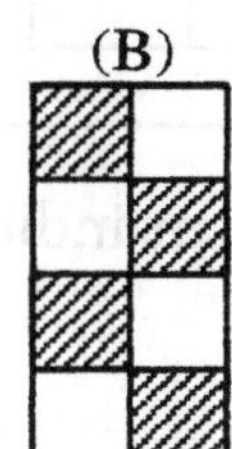
(C) 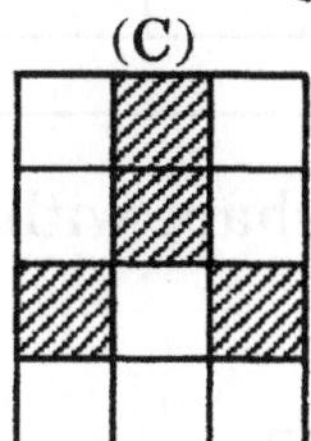
(D) 

12 If Alex wrote down 19.3 instead of 1.93 in a problem, how much more than the original number did he put down?

(A) 17.37 (B) 7.37
(C) 1.37 (D) 0.37

13 The most likely missing number in the following series would be :

$$\frac{1}{3}, \frac{1}{6}, \frac{1}{12}, \square, \frac{1}{48}$$

(A) $\frac{1}{32}$ (B) $\frac{1}{24}$
(C) $\frac{1}{9}$ (D) $\frac{1}{16}$

14 Four types of pens are sold in a stationery shop. Which type is the cheapest?

(A) \$4 for 3
(B) \$10 per half dozen
(C) \$8 for 5
(D) \$1.20 each

15 23% of the people in a room are women, 45% of them are men and the rest are children. What fraction of the people are children?

(A) $\frac{8}{25}$ (B) $\frac{23}{50}$
(C) $\frac{11}{20}$ (D) $\frac{23}{100}$

16 51 out of 68 workers in a department are female. What percentage of the workers are male?

(A) 75% (B) 50%
(C) 25% (D) 40%

17 Which of the given shapes has turning symmetry through 180° about the point O?

(P) 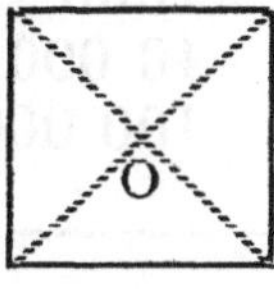

(Q) 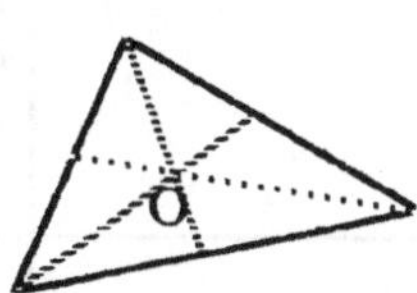

(R) 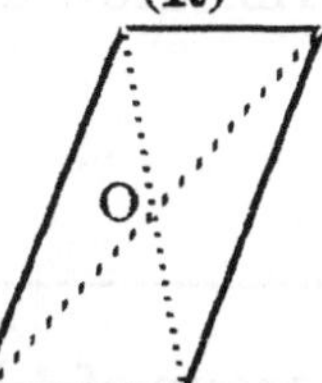

(S) 

(A) (P) only (B) (P) and (Q) only
(C) (P) and (R) only (D) (S) and (S) only

18 Using the subdivisions on the bar chart given, what percentage of the parents of Primary 6D pupils are public servants? Occupations of parents of Primary 6D pupils Public servants | Merchants | Clerks | Other	(A) 25% (B) 35% (C) 40% (D) 50%

19 Match the cross-sectional shape with the cut indicated on the cube.

(A) (B) (C) (D)

20 How long will 2 tonnes 430kg of firewood last if a stove burns 30kg of wood each day?	(A) 27 days (B) 810 days (C) 81 days (D) None of these
21 If eight hundred and twelve thousand three hundred and twenty eight is subtracted from one million two hundred and eight thousand one hundred and seven, the result is	(A) 604 221 (B) 395 779 (C) 467 779 (D) 395 842
22 Arrange in ascending order of length: 47mm, 4.8cm, 0.46m, 0.00045km	(A) 0.00045km, 0.46m, 47mm, 4.8cm (B) 47mm, 4.8cm, 0.00045km, 0.46m (C) 0.46m, 0.00045km, 4.8cm, 47mm (D) 0.00045km, 0.46m, 4.8cm, 47mm
23 The distance between two places of interest is $\frac{2}{3}$ km. I go $\frac{1}{4}$ of this distance on foot. How far is still to go?	(A) $\frac{1}{6}$ km (B) $\frac{1}{4}$ km (C) $\frac{1}{2}$ km (D) $\frac{5}{6}$ km
24 The side of a square is 1m. How many cm^2 is its area?	(A) 100 (B) 1000 (C) 10 000 (D) 100 000
25 The average mass of a group of 4 boys is 80kg. Another boy of mass 70kg joins the group. What is the average mass of the group now?	(A) 75kg (B) 78kg (C) 64kg (D) 76kg

26	If 8% of the water in a tank is 56 litres, how much water is in the tank?	(A) 448 litres (B) 7 litres (C) 44.8 litres (D) 700 litres
27	A rectangular paddock is of area $1\frac{1}{2}$ ha. If its length is 300m then its perimeter is	(A) 50m (B) 350m (C) 700m (D) 1 600m
28	A flight took from 2250h Tuesday to 0815h Thursday of the same week. How long was the flight?	(A) 9h 25min (B) 33h 25min (C) 34h 5min (D) 10h 25min

29 - 33 The solid shown is made up of equal cubes. The 6 faces are painted blue, and the solid broken up into individual cubes.

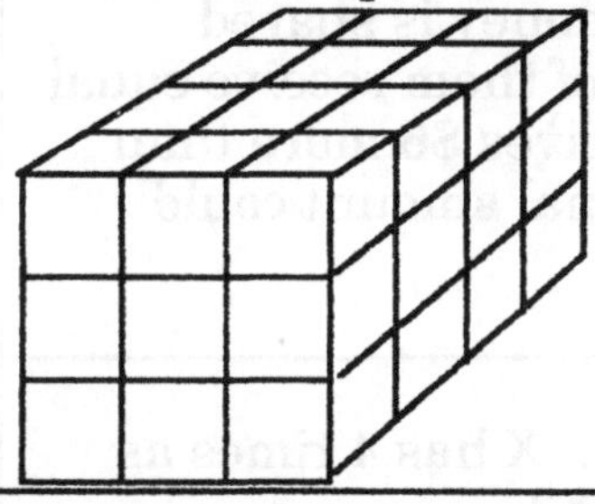

29	How many distinct cubes are not visible in the sketch, (before the breakup) ?	(A) 12 (B) 13 (C) 3 (D) 11
30	After the break up, how many cubes have exactly one face painted blue?	(A) 6 (B) 8 (C) 10 (D) 12
31	How many individual cubes have exactly two faces painted blue?	(A) 10 (B) 12 (C) 14 (D) 16
32	How many cubes have exactly three faces painted blue?	(A) 6 (B) 8 (C) 10 (D) 12
33	How many cubes have no faces painted blue?	(A) 0 (B) 1 (C) 2 (D) 4

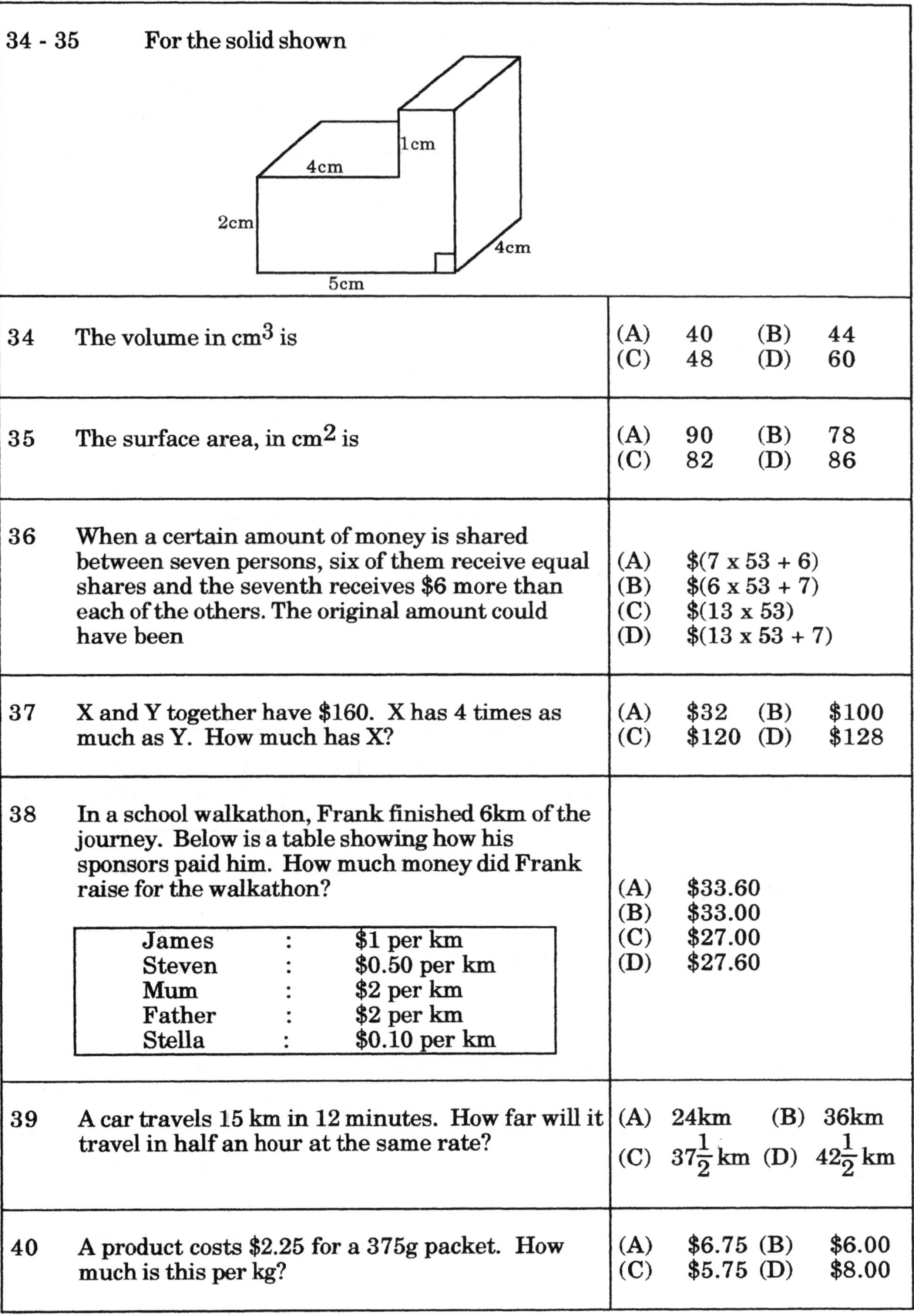

34 - 35 For the solid shown

34	The volume in cm^3 is	(A) 40 (B) 44 (C) 48 (D) 60
35	The surface area, in cm^2 is	(A) 90 (B) 78 (C) 82 (D) 86
36	When a certain amount of money is shared between seven persons, six of them receive equal shares and the seventh receives \$6 more than each of the others. The original amount could have been	(A) \$(7 x 53 + 6) (B) \$(6 x 53 + 7) (C) \$(13 x 53) (D) \$(13 x 53 + 7)
37	X and Y together have \$160. X has 4 times as much as Y. How much has X?	(A) \$32 (B) \$100 (C) \$120 (D) \$128
38	In a school walkathon, Frank finished 6km of the journey. Below is a table showing how his sponsors paid him. How much money did Frank raise for the walkathon? James : \$1 per km Steven : \$0.50 per km Mum : \$2 per km Father : \$2 per km Stella : \$0.10 per km	(A) \$33.60 (B) \$33.00 (C) \$27.00 (D) \$27.60
39	A car travels 15 km in 12 minutes. How far will it travel in half an hour at the same rate?	(A) 24km (B) 36km (C) $37\frac{1}{2}$ km (D) $42\frac{1}{2}$ km
40	A product costs \$2.25 for a 375g packet. How much is this per kg?	(A) \$6.75 (B) \$6.00 (C) \$5.75 (D) \$8.00

PAPER 6

1	The digit 6 in the number 94.68 is changed to 2. What is the difference between the original number and the new number?	(A) 10 (B) 40 (C) 20 (D) 0.4
2	$2\frac{1}{2} \div 1\frac{3}{4} =$	(A) $1\frac{3}{7}$ (B) $2\frac{2}{3}$ (C) $4\frac{3}{8}$ (D) $\frac{7}{10}$
3	Express 3.45 as a mixed number.	(A) $3\frac{9}{20}$ (B) $3\frac{9}{100}$ (C) $3\frac{9}{50}$ (D) $3\frac{9}{25}$
4	What is the difference between 3 right angles and $4\frac{1}{2}$ right angles?	(A) 240° (B) 180° (C) 150° (D) 135°
5	Which one of the following figures has both parallel sides and perpendicular sides? (A) (B) (C) (D)	
6	What is the lowest common multiple of 4, 9 and 10	(A) 180 (B) 40 (C) 360 (D) 90
7	When 29 is divided by 6, the quotient and remainder are respectively	(A) 5 and 1 (B) 4 and 5 (C) 3 and 7 (D) 5 and 4

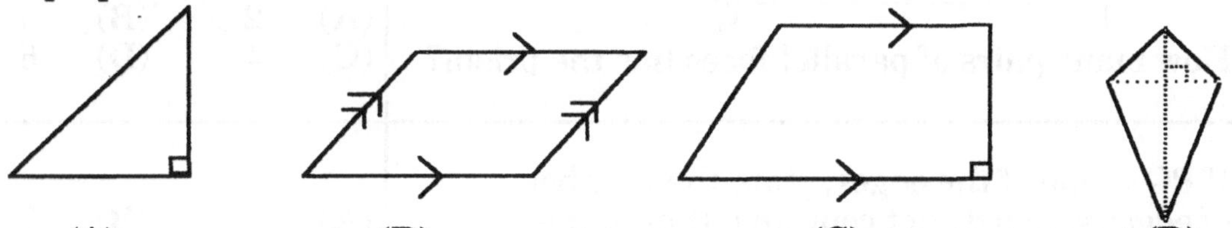

8 - 9 The sketch shows a stack of unit cubes. 8 What is the volume of the stack (in cubic units)?	(A) 10 (B) 16 (C) 20 (D) 24
9 What is the surface area visible in the sketch (in square units)?	(A) 10 (B) 20 (C) 40 (D) 30
10 - 11 The sketch shows a rectangular prism. P Q 10 How many pairs of parallel faces has the prism?	(A) 2 (B) 3 (C) 4 (D) 6
11 If PQ is one of the edges, state the number of edges which do not contain either of the points P, Q.	(A) 8 (B) 7 (C) 5 (D) 6
12 At a theatre, a child's ticket costs \$6. Mr. and Mrs. Psaltis and their two children paid a total of \$35. How much did each adult's ticket cost?	(A) \$11.50 (B) \$10.50 (C) \$9.50 (D) \$8.50
13 Express 12 minutes as a percentage of 2 hours.	(A) 10% (B) 20% (C) 40% (D) 50%
14 Which of the following is the largest?	(A) 30% of 14m (B) 60% of 6m (C) 80% of 5m (D) 50% of 9m

15 - 16 In the following table :

Shapes	Number of edges (E)	Number of vertices (V)	Number of faces (F)	E + V + F
Pentagonal Prism			(P)	
Hexagonal Pyramid				(Q)

15 What is the value of P?

(A) 5 (B) 6
(C) 7 (D) 8

16 What is the value of Q?

(A) 26 (B) 19
(C) 14 (D) 12

17 - 20 The graph given shows the journey of a motorist from his home to the city and back. Use it to answer the questions below.

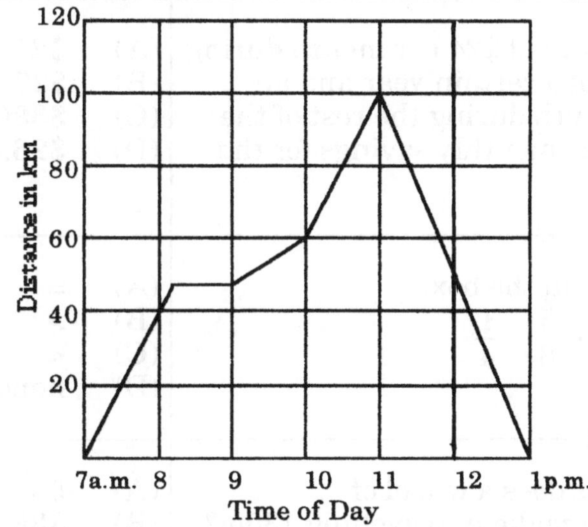

17 At what average speed did the motorist travel during the first hour?

(A) 20km/h
(B) 30km/h
(C) 40km/h
(D) 50km/h

18 What was his average speed for the return journey?

(A) 40 km/h
(B) 45km/h
(C) 50km/h
(D) 55km/h

19 How far from the city was he at 10a.m?

(A) 60km (B) 50km
(C) 80km (D) 40km

20	The second time that the motorist was 40km from home was approximately	(A) 8a.m. (B) 11.45a.m. (C) 12.15a.m. (D) 12.15p.m.
21	Jeff bought a car for $42 500. He paid a deposit and a monthly instalment of $950 for 3 years in full payment. How much was his deposit?	(A) $7300 (B) $8300 (C) $9300 (D) $10 300
22	Which of the following is closest to $\frac{3}{20}$?	(A) $\frac{9}{40}$ (B) $\frac{21}{100}$ (C) $\frac{1}{4}$ (D) $\frac{1}{5}$
23	Angela saved an average of $24 per month during the first five months of a certain year and an average of $30 per month during the rest of the year. Find her average monthly savings for the whole year.	(A) $27 (B) $27.50 (C) $330 (D) $26.50
24	Insert the correct sign in the box. $\frac{7}{14} - \frac{2}{7} + \frac{11}{28} \ \square \ \frac{3}{8} + \frac{5}{8} - \frac{1}{4}$	(A) = (B) > (C) < (D) none of these
25	How many revolutions does a wheel of circumference of 2.6m make in travelling 130m?	(A) 50 (B) 338 (C) 0.02 (D) None of these
26	All the divisors of 36, including 36 itself, are added together. Their sum is	(A) 90 (B) 55 (C) 91 (D) 37
27	3.57m is equal to	(A) 35.7cm (B) 357cm (C) 0.357cm (D) 0.0357cm
28	Twelve year old Jody used her handspan to measure the length of a table. The table was 15 handspans long. The length of the table was approximately	(A) 1.5m (B) 2.4m (C) 7.2m (D) 10.0m

29	The side of a cube is 1m. How many cm^3 is it in volume?	(A) 100 (B) 10 000 (C) 1 000 000 (D) 100 000 000
30	The 20th century runs from the year 1901 to the year 2000, both inclusive. How many months is this?	(A) 1188 (B) 1200 (C) 1212 (D) 1560
31	If $2\frac{1}{2}$ equal lengths of cloth measure 8m, what will $1\frac{1}{2}$ such lengths measure?	(A) 3.2m (B) 4m (C) 4.8m (D) 5.4m
32	In the sketch, X is the mid-point of the side SR of the rectangle shown. What is the size of the shaded area? 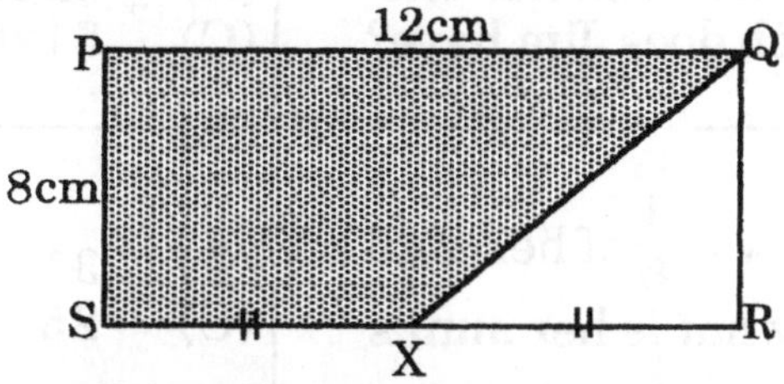 12cm, P, Q, 8cm, S, X, R	(A) $24cm^2$ (B) $72cm^2$ (C) $84cm^2$ (D) $48cm^2$
33	The perimeter of the isosceles triangle ABC shown in the sketch is 42cm. If BC is longer than AB or AC, then BC could be 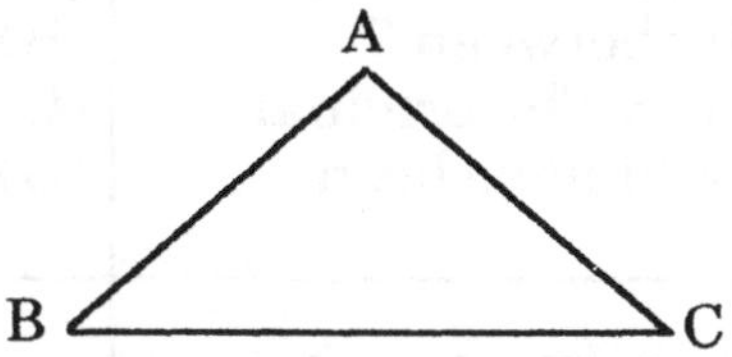 A, B, C	(A) 12cm (B) 14cm (C) 16cm (D) any of these
34	The sketch shows a sector graph, in which a central angle is 72°. What percentage of the graph is shaded? 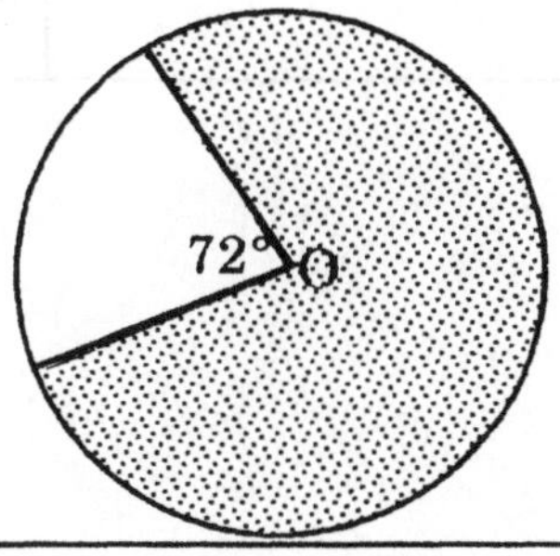 72°, O	(A) 20% (B) 80% (C) 28% (D) 75%

35 - 36 The sketch shows the base of a prism of height 12cm. [Diagram labels: 3cm, 2cm, 5cm, 5cm, 6cm] 35 The area of the base is	(A) $94cm^2$ (B) $58cm^2$ (C) $70cm^2$ (D) $38cm^2$
36 The volume of the prism is	(A) $106cm^3$ (B) $840cm^3$ (C) $82cm^3$ (D) $1128cm^3$
37 Bill and Jim together have \$320. Jim has one third of Bill's share. How much does Jim have?	(A) \$240 (B) \$160 (C) \$120 (D) \$80
38 Seven years ago, Gloria's age was $\frac{1}{6}$ of her aunt's. If Gloria is 11 years old now, what is her aunt's present age in years?	(A) 31 (B) 27 (C) 25 (D) 22
39 If a certain amount of money is shared equally between 6 persons, there is \$1 left over. If the same amount is shared equally between 7 persons, there is also \$1 left over. The original amount of money in dollars could have been	(A) 6 x 13 x 40 + 1 (B) 13 x 49 x 19 + 1 (C) 6 x 49 x 5 + 1 (D) 36 x 53 x 13 + 1
40 A rectangle has perimeter 24cm. The length is twice the breadth. What is the area of the rectangle?	(A) $32cm^2$ (B) $128cm^2$ (C) $8cm^2$ (D) $48cm^2$

PAPER 7

1	In 162.57, the digit 7 stands for	(A) 7 ones (B) 7 tenths (C) 7 hundredths (D) 7 tens
2	12.74 = 10 + 2 + ❑ + 0.04. The missing number in the box is	(A) 70 (B) 7 (C) 0.7 (D) 0.07
3	Express 40% as a fraction.	(A) $\frac{1}{40}$ (B) $\frac{2}{5}$ (C) $\frac{1}{4}$ (D) $\frac{3}{5}$
4	What percentage is 7mL of 1L?	(A) 7% (B) 70% (C) 0.7% (D) 0.07%
5	The area of a square is 1cm^2. What is the length of a side in mm?	(A) 1mm (B) 10mm (C) 0.1mm (D) 100mm
6	The number of kg in 654 300 g is	(A) 6.543 (B) 65.43 (C) 654.3 (D) 6543
7	A vehicle weighs 5.1t when loaded and 2.7t when unloaded. What is the weight of the load?	(A) 2400kg (B) 7800kg (C) 240kg (D) 780kg
8	A rectangular carpet 4m x 3.5m is on the floor of a room 6m x 4.5m. The uncovered area is	(A) 41m^2 (B) 13m^2 (C) 12m^2 (D) 14m^2

9 - 11 The graph shown is used to convert degrees Celsius (°C) to degrees Fahrenheit (°F), and vice versa. Use the graph to answer the following questions.

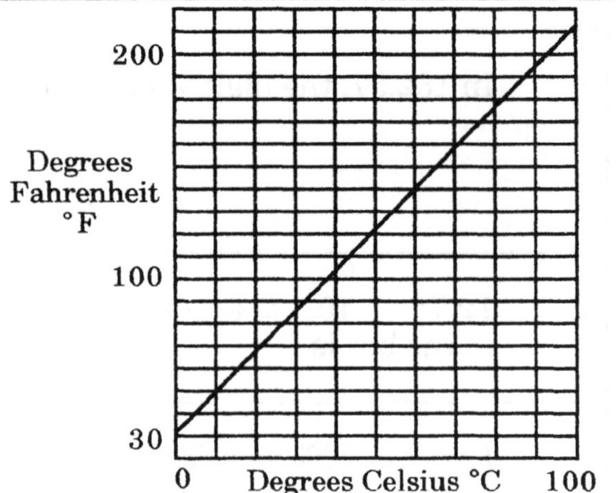

9	The approximate freezing point of water and the approximate boiling point of water in °F are respectively	(A) 0°F, 100°F (B) 32°F, 180°F (C) 32°F, 212°F (D) 0°F, 212°F
10	60°C is approximately equal to	(A) 136°F (B) 144°F (C) 130°F (D) 140°F
11	68°F is approximately equal to	(A) 20°C (B) 18°C (C) 22°C (D) 24°C
12	Which one of the following is incorrect?	(A) $\frac{1}{3} < 0.5$ (B) 4 tenths more than 0.35 is 0.75 (C) $6.25 = 6 + \frac{2}{10} + \frac{5}{100}$ (D) 0.22 > 2 tenths 2 hundredths
13	When 67 is divided by a certain number, the quotient is 22 and the remainder is 1. The number must have been	(A) 45 (B) 5 (C) 3 (D) None of these

14 The data below shows the cost price to a retailer of different sized jars of jam.

Size	Price
Small	$1.55
Medium	$2.80
Large	$4.65

The retailer bought a dozen small jars of jam and half a dozen large jars of jam. How much would these cost her altogether?

(A) $46.50 (B) $36.50
(C) $28.05 (D) $34.55

15	The price of a pen is 6 times that of a pencil. If the number of pens sold today was $\frac{2}{3}$ of the number of pencils sold then $\frac{\text{the number of pens sold}}{\text{the number of pencils sold}} = \ldots\ldots$	(A) $\frac{2}{3}$ (B) $\frac{1}{3}$ (C) $\frac{3}{2}$ (D) $\frac{1}{6}$
16	The thickness of 100 sheets of paper is 2.5cm. Find the thickness of 2 sheets of paper.	(A) 25mm (B) 2.5mm (C) 5mm (D) 0.5mm
17	70% of 1.2km is	(A) 84m (B) 480m (C) 140m (D) 840m
18	There are 50 pupils in a group. 20% of them play cricket. How many pupils in the group do not play cricket?	(A) 1000 (B) 40 (C) 10 (D) 25

19 Which of the following graphs shows a relationship between P and Q such that, as P increases, Q decreases, and vice versa?

Q, P (A) Q, P (B) Q, P (C) Q, P (D)

20	The 4-digit number 56 □ 1 is divisible by 23. The number in the box must be	(A) 5 (B) 6 (C) 3 (D) 8
21	On a number line, which number is three times as far from 9 as 7 is?	(A) 15 only (B) 3 and 15 (C) 6 only (D) 6 and 15
22	A woman earns \$52 000 each year. If she saves \$225 each week, how much does she spend each year? [Take the year as having 52 weeks.]	(A) \$11 700 (B) \$40 300 (C) \$32 400 (D) \$42 600

23	Apart from itself and 1, certain numbers have exactly 3 divisors. Of the following numbers, the smallest such number would be	(A) 8 (B) 12 (C) 16 (D) 25
24	One hectare of land is being subdivided into blocks measuring 25 metres by 20 metres. The number of such blocks will be	(A) 40 (B) 10 (C) 20 (D) 60
25	Thirty kilograms of copper are being made into copper wire such that 40 metres of wire weigh 1kg 200g. How many metres of wire can be made?	(A) 1000 (B) 1200 (C) 2000 (D) 2400
26	A girl buys 16m of cloth for \$6. She sells 10m of it for \$1.30 per metre. For how much per metre must she sell the remainder if she wants to make a total profit of \$16?	(A) \$1.50 (B) \$9 (C) \$1.42 (D) \$1.66
27	How many metres of silk at \$4.20 per metre can be bought with \$29.40?	(A) 9 (B) 7 (C) 5 (D) None of these
28	How many multiples of 3 are there between 103 and 154?	(A) 17 (B) 16 (C) 18 (D) 20
29	The area of each face of a cube is $16cm^2$. The volume of the cube is	(A) $12cm^3$ (B) $16cm^3$ (C) $64cm^3$ (D) $256cm^3$
30	The outside dimensions of a closed rectangular empty wooden box are 6cm x 5cm x 4cm. If the wood is 5mm thick, what are the inside dimensions of the box?	(A) 5.5cm x 4.5cm x 3.5cm (B) 5cm x 4cm x 3cm (C) 6.5cm x 5.5cm x 4.5cm (D) 7cm x 6cm x 5cm
31	The time for a satellite to orbit the Earth is 93.6 minutes. How many complete orbits will the satellite make in 9 hours?	(A) 72 (B) 5 (C) 6 (D) 10

32 - 33 Signals travel in our nervous system at the rate of 14 m/s.	
32 How far would a nervous system signal travel in 1 hour?	(A) 840m (B) 50.4km (C) 20.16km (D) 3.6km
33 How long would it take for such a signal to travel 21km?	(A) 25min (B) 62.5h (C) 2.5min (D) 30min
34 - 35 All the faces of the cube shown are marked with odd numbers. 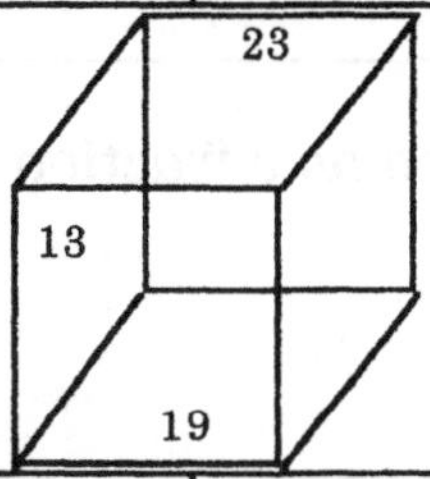	
34 What is the largest possible number on the cube if the numbers are consecutive odd numbers?	(A) 25 (B) 23 (C) 27 (D) 21
35 What is the largest possible number on the cube if the numbers are consecutive prime numbers?	(A) 37 (B) 27 (C) 29 (D) 31
36 The sketch shows the net of a 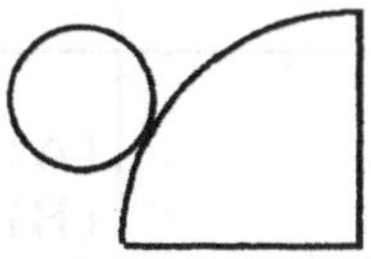	(A) cylinder (B) tetrahedron (C) cone (D) none of these
37 How many squares are there in the figure given? 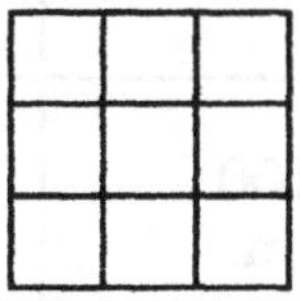	(A) 9 (B) 12 (C) 14 (D) 16
38 For every \$10 that I save, my father gives me \$4. How much must I save if I want to have \$84?	(A) \$80 (B) \$40 (C) \$50 (D) \$60
39 Tom, Shmuel and Jan shared \$72. Tom received three times as much as Shmuel who received the same amount as Jan. How much money did Jan receive?	(A) \$18 (B) \$54 (C) \$14.40 (D) \$10
40 A class of 25 students decide to put in \$1.20 each in order to buy a certain present for their teacher. If 5 students forget to bring the money, how much more will the remaining students have to contribute so as to buy this present?	(A) \$1.50 (B) \$0.80 (C) \$0.40 (D) \$0.30

FURTHER MATHEMATICS TESTS

PAPER 8

1	Which one of the following is the best estimate for 15.8 x 7.59 ?	(A) 16 x 8 (B) 16 x 7 (C) 15 x 7 (D) 15 x 8
2	$2\frac{7}{20}$ expressed as a decimal is	(A) 2.7 (B) 2.07 (C) 2.35 (D) 2.05
3	0.4 expressed as a fraction in its lowest terms is	(A) $\frac{4}{10}$ (B) $\frac{4}{100}$ (C) $\frac{2}{5}$ (D) $\frac{2}{25}$
4	By how much does five hundred and four exceed three hundred and nine?	(A) 155 (B) 231 (C) 114 (D) None of these
5	The product of the sum of 3 and 5 and the difference of 11 and 2 is	(A) 72 (B) 36 (C) 17 (D) 330
6	The number of m in 2km 70m 4cm is	(A) 2070.04 (B) 2070.4 (C) 274 (D) 270.4
7	The scale of the floor plan of a house is 1:100. The length of a hallway in the plan is 75mm. What is the actual length of the hallway?	(A) 7.5km (B) 7.5m (C) 7.5cm (D) 75m
8	If $\frac{2}{3}$ of a number is 24, then $\frac{3}{4}$ of the same number is	(A) 12 (B) 48 (C) $21\frac{1}{3}$ (D) 27

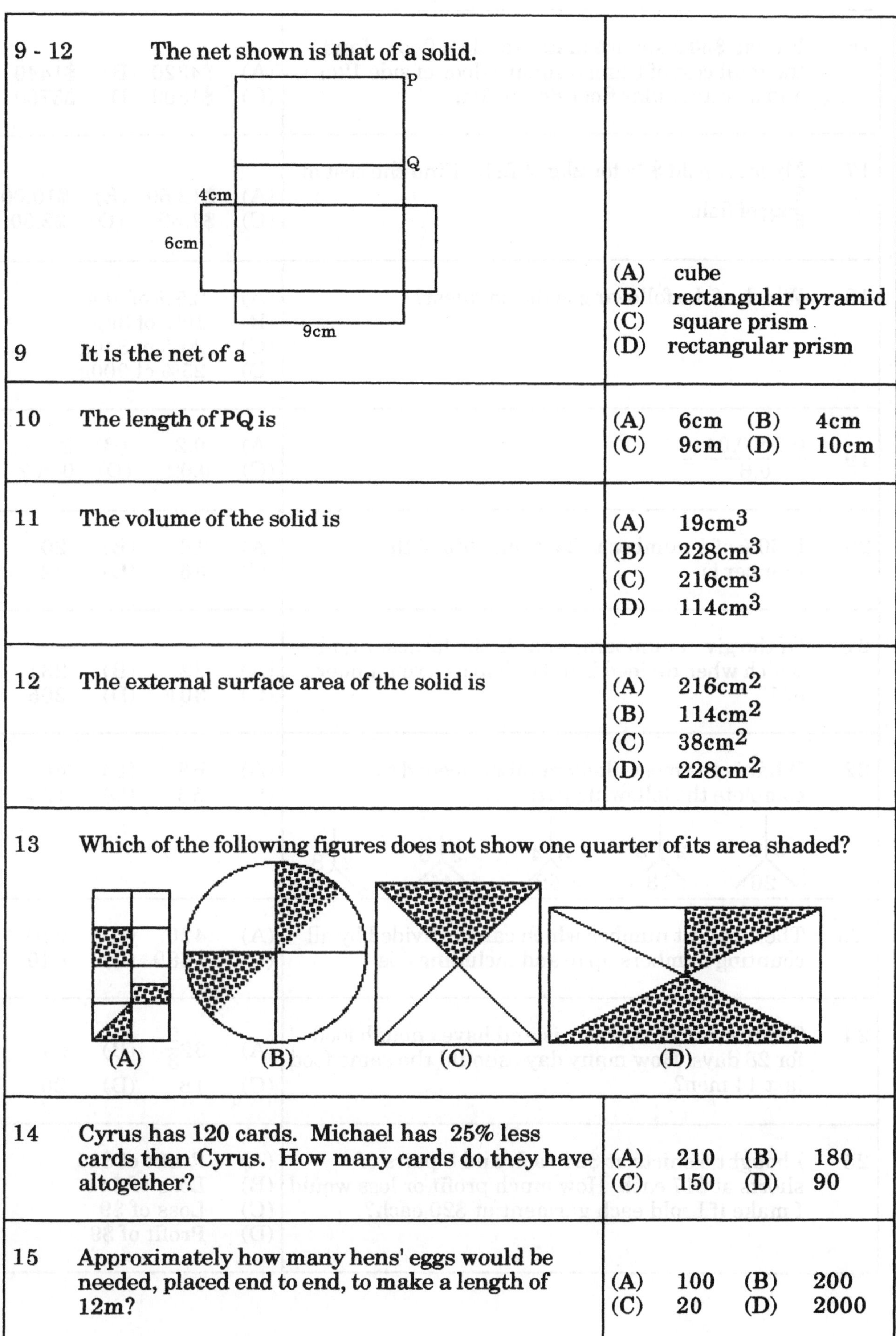

9 - 12 The net shown is that of a solid.

9 It is the net of a

(A) cube
(B) rectangular pyramid
(C) square prism
(D) rectangular prism

10 The length of PQ is

(A) 6cm (B) 4cm
(C) 9cm (D) 10cm

11 The volume of the solid is

(A) $19cm^3$
(B) $228cm^3$
(C) $216cm^3$
(D) $114cm^3$

12 The external surface area of the solid is

(A) $216cm^2$
(B) $114cm^2$
(C) $38cm^2$
(D) $228cm^2$

13 Which of the following figures does not show one quarter of its area shaded?

(A) (B) (C) (D)

14 Cyrus has 120 cards. Michael has 25% less cards than Cyrus. How many cards do they have altogether?

(A) 210 (B) 180
(C) 150 (D) 90

15 Approximately how many hens' eggs would be needed, placed end to end, to make a length of 12m?

(A) 100 (B) 200
(C) 20 (D) 2000

16	It costs \$30 a square metre to tile a floor. Find the total cost of tiling a square floor of side 12m and a rectangular floor 8m by 6m.	(A) \$4320 (B) \$1440 (C) \$1800 (D) \$5760
17	My aunt paid \$25 for 2kg of fish. Find the cost of $\frac{3}{5}$ kg of fish.	(A) \$12.50 (B) \$10.00 (C) \$7.50 (D) \$2.50
18	Which of the following is the smallest?	(A) 2.5% of 4kg (B) 10% of 3kg (C) 20% of 840g (D) 25% of 900g
19	$\frac{0.3 \times 0.04}{0.6} =$	(A) 0.2 (B) 2 (C) 0.02 (D) 0.002
20	If 30% of a number is 24, then 40% of the number is	(A) 18 (B) 20 (C) 36 (D) 32
21	Of the given numbers, what is the largest number which when divided into 411 leaves a remainder of 1?	(A) 82 (B) 231 (C) 301 (D) 205
22	What is the most likely number needed to complete the following pattern? 5 1 / 26 2 3 / 13 6 4 / 52 3 6 / 45 2 8 /	(A) 68 (B) 48 (C) 56 (D) 124
23	The smallest number which can be divided by all counting numbers up to and including 8 is	(A) 420 (B) 240 (C) 1680 (D) 840
24	Twelve men on a desert island have enough food for 28 days. How many days should the same food last 14 men?	(A) $32\frac{2}{3}$ (B) 24 (C) 18 (D) 20
25	I bought 5 shirts at \$23 each and 8 pairs of shorts at \$17 each. How much profit or loss would I make if I sold each garment at \$20 each?	(A) Profit of \$15 (B) Loss of \$4 (C) Loss of \$9 (D) Profit of \$9

26 A girl is rowing downstream from one town to another town 14km away. If the stream is flowing at 4km/h and the girl can row at 3km/h in still water, how long will she take?	(A) 14 hours (B) 7 hours (C) $3\frac{1}{2}$ hours (D) None of these
27 - 28 A room is 5m x 4m x $3\frac{1}{2}$ m 27 Ignoring windows and doors, the area of the 4 walls of the room is	(A) $51.5m^2$ (B) $103m^2$ (C) $126m^2$ (D) $63m^2$
28 The capacity of the room is	(A) $103m^3$ (B) $70m^3$ (C) $25m^3$ (D) $84m^3$
29 0.75 of a kg in g is	(A) 75 (B) 7.5 (C) 750 (D) 7500
30 A car travels 480km between 9 a.m. and 4:30 p.m.. What is the average speed in km/h?	(A) 60km/h (B) 64km/h (C) 56km/h (D) 72km/h
31 The masses of 2 objects are 5400g and 0.0146t. What is their total mass in kg?	(A) 20kg (B) 68.6kg (C) 2kg (D) 200kg
32 The lap of a race track is 400m. How many laps are required for a 100km race?	(A) 25 (B) 250 (C) 2500 (D) 320
33 - 34 A rectangular garden 11m x 8m is surrounded by a path 1m wide. 33 The outside dimensions of the garden-path area are	(A) 12m x 9m (B) 10m x 7m (C) 13m x 10m (D) 9m x 6m
34 The area of the path is	(A) $42m^2$ (B) $22m^2$ (C) $60m^2$ (D) $76m^2$

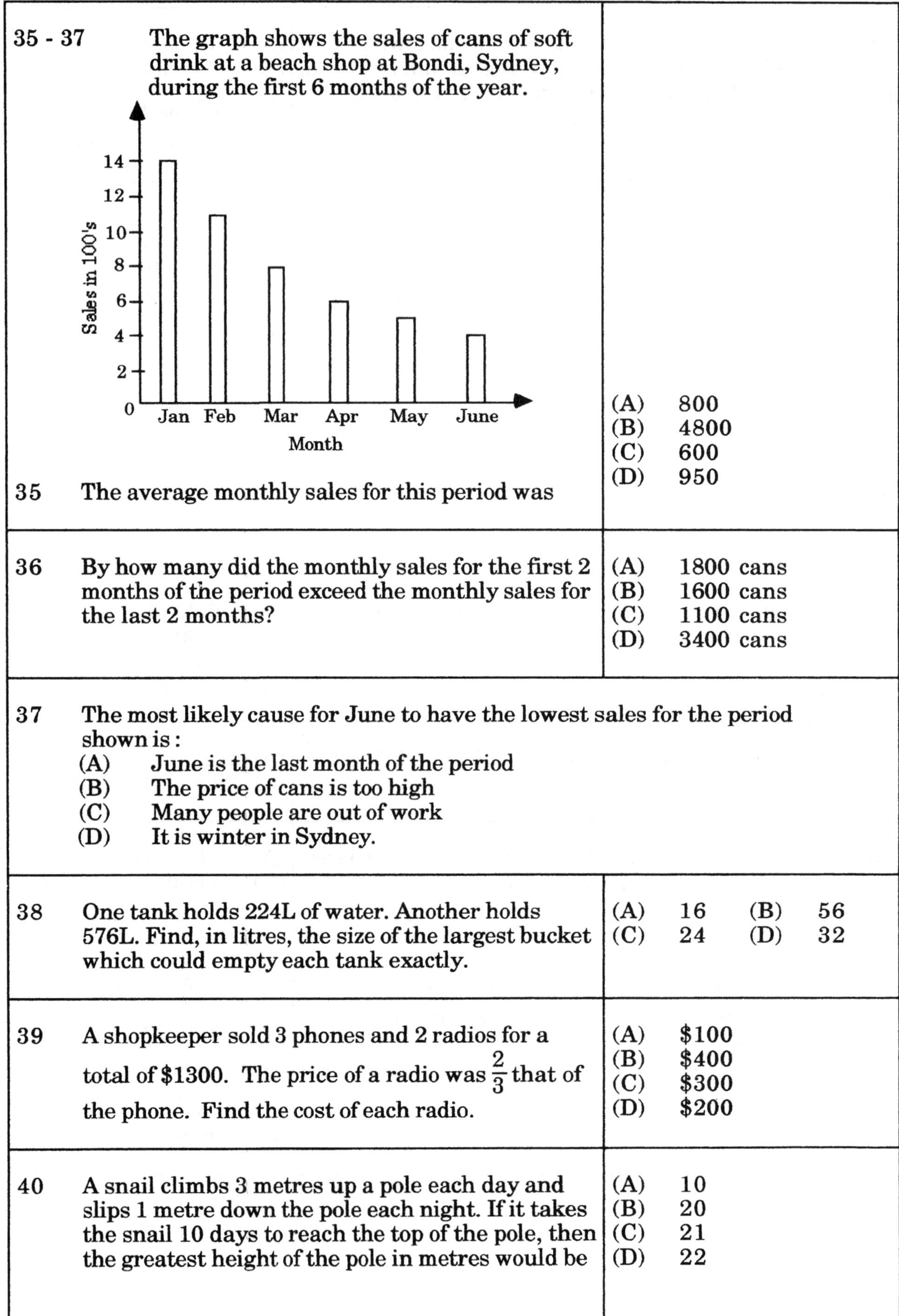

35 - 37 The graph shows the sales of cans of soft drink at a beach shop at Bondi, Sydney, during the first 6 months of the year.

35 The average monthly sales for this period was

(A) 800
(B) 4800
(C) 600
(D) 950

36 By how many did the monthly sales for the first 2 months of the period exceed the monthly sales for the last 2 months?

(A) 1800 cans
(B) 1600 cans
(C) 1100 cans
(D) 3400 cans

37 The most likely cause for June to have the lowest sales for the period shown is :

(A) June is the last month of the period
(B) The price of cans is too high
(C) Many people are out of work
(D) It is winter in Sydney.

38 One tank holds 224L of water. Another holds 576L. Find, in litres, the size of the largest bucket which could empty each tank exactly.

(A) 16 (B) 56
(C) 24 (D) 32

39 A shopkeeper sold 3 phones and 2 radios for a total of \$1300. The price of a radio was $\frac{2}{3}$ that of the phone. Find the cost of each radio.

(A) \$100
(B) \$400
(C) \$300
(D) \$200

40 A snail climbs 3 metres up a pole each day and slips 1 metre down the pole each night. If it takes the snail 10 days to reach the top of the pole, then the greatest height of the pole in metres would be

(A) 10
(B) 20
(C) 21
(D) 22

PAPER 9

1	Divide 68.8 by 20. The answer is	(A) 0.344 (B) 3.44 (C) 34.4 (D) 344
2	9kg 600g = ... kg	(A) 9.006 (B) 9.06 (C) 9.6 (D) 96
3	3% of 6m is	(A) 18m (B) 18cm (C) 18mm (D) 2cm
4	How many 25kg bags make up 2 tonnes?	(A) 8 (B) 80 (C) 800 (D) 8000
5	The usual temperature inside a refrigerator is approximately	(A) 0°C (B) 5°C (C) 12°C (D) 10° below zero
6	The number of axes of symmetry for this figure is	(A) 4 (B) 6 (C) 8 (D) 12
7	Alfredo is 1.61m in height whilst his sister Anna is 1540mm in height. How much shorter is Anna than Alfredo?	(A) 17cm (B) 7cm (C) 70cm (D) 11cm
8	Find the difference between 4% of \$50 and 6% of \$48.	(A) \$3.88 (B) \$2.88 (C) \$1.88 (D) \$0.88
9	Subtract 4.68 from the product of 2.49 and 7. The answer is	(A) 17.43 (B) 12.75 (C) 22.11 (D) 2.32
10	What is the most likely number needed to complete the following pattern? 3, 4, 7, 11, 18, 29, ☐	(A) 37 (B) 47 (C) 49 (D) 51
11	A group of forty people went to a show. An adult's ticket costs \$12 and a child's ticket costs \$6. If three tenths of the group were children, how much did the group pay altogether?	(A) \$408 (B) \$480 (C) \$720 (D) \$144

12 There are five cards with numbers written on them as shown below. [2] [5] [6] [7] [9] Two of them are drawn at a time and put together to form fractions, e.g. $\frac{\boxed{2}}{\boxed{5}}$ How many proper fractions can be formed?	(A) 2 (B) 4 (C) 7 (D) 10
13 The line PQ is perpendicular to the line RS and RS is perpendicular to the line TU. Therefore	(A) PQ is perpendicular to TU (B) PQ is parallel to TU (C) PQ = TU (D) None of these
14 The average cost of 8 calculators in a particular shop is \$32.50. If three of them cost \$29 each, what is the average cost of the other five?	(A) \$34.60 (B) \$32.50 (C) \$30.50 (D) \$29
15 The 4-digit number 41 □ 3 is divisible by 3. The number in the box must be	(A) 1 only (B) 3 only (C) 1 or 4 only (D) 1 or 4 or 7 only
16 A wheel makes exactly 60 revolutions in travelling 72 metres. The circumference of the wheel is	(A) 85cm (B) 1.2m (C) 12cm (D) 4.32m
17 After a 10% discount, a skirt is sold at a price of \$54. What is the original price of 10 such skirts?	(A) \$500 (B) \$700 (C) \$450 (D) \$600
18 The greater of two numbers is 4127 and their sum is 7003. The other number is	(A) 2876 (B) 11 130 (C) 2903 (D) 3173
19 If 12kg of premium coffee cost \$42, the cost of 17kg will be	(A) \$29.65 (B) \$61 (C) \$59.50 (D) \$57.50
20 If 5 plangs = 12 plings and 3 plings = 8 plungs, how many plangs in 160 plungs?	(A) $3\frac{1}{3}$ (B) $19\frac{1}{5}$ (C) 36 (D) 25

21	The product of two numbers is 24 and their quotient is $\frac{2}{3}$. The smaller of the two numbers is	(A) 4 (B) 6 (C) 2 (D) None of these
22	Which of the following numbers leaves a remainder of 2 when divided by each of 7, 11 and 13?	(A) 7 x 11 x 13 + 2 (B) 7 x 11 x 2 + 13 (C) 11 + 7 x 13 x 2 (D) 2 x 11 x 13 + 7
23	The perimeter of a square is 1m. What is its area?	(A) $6.25cm^2$ (B) $\frac{1}{4}m^2$ (C) $\frac{1}{2}m^2$ (D) $625cm^2$
24	How many m^2 in 1 km^2?	(A) One thousand (B) One million (C) One billion (D) Ten thousand
25	The Olympic Games are staged each leap year. They were first held in 776 B.C. If they had occurred continuously since then, in which of the following years should they have been held? 614 920 B.C. 1870 1908	(A) 614 and 1908 (B) 920 B.C. only (C) 1908 only (D) 920 B.C. and 1908
26	The volume in cm^3 of a rectangular box 8cm x 4.5cm x 3.5cm is	(A) 96 (B) 126 (C) 108 (D) 162
27	A train passes through a station at 10:06 a.m. travelling at a speed of 80km/h. At what time should it pass through the next station 32km away?	(A) 10:30 a.m. (B) 10:24 a.m. (C) 10:54 a.m. (D) 11 a.m.
28	A rectangular farm is 6ha in area. If its breadth is 120m, what is its length?	(A) 500m (B) 50m (C) 5000m (D) 720m

29 An athlete covered the same distance daily in running exercises. The first run took $48\frac{3}{4}$ minutes and a week later the run took $43\frac{1}{2}$ minutes. What was the average daily decrease in time for the run?	(A) 30s (B) 45s (C) 60s (D) 75s
30 - 32 The bar graph, not drawn to scale, shows the distribution of weekly income for a certain couple. Rent (3cm) | Food (4cm) | Clothing (2cm) | Misc. (1cm) 30 What percentage of their income is spent on Food?	(A) $33\frac{1}{3}\%$ (B) 40% (C) 50% (D) 45%
31 If their total income for one particular week was \$750, how much was spent on Clothing in that week?	(A) \$150 (B) \$200 (C) \$225 (D) \$175
32 For another particular week, the couple found that the amount spent on Food exceeded the amount spent on Miscellaneous by \$216. What was their income for that week?	(A) \$648 (B) \$432 (C) \$720 (D) \$840
33 - 34 The sketch shows centimetre cubes, which are being stacked to form a prism 4cm x 3cm x 2cm. 33 How many more cubes are required?	(A) 24 (B) 17 (C) 18 (D) 16
34 What is the surface area of the stack shown in the sketch (do not include the base) ?	(A) $18cm^2$ (B) $17cm^2$ (C) $19cm^2$ (D) $20cm^2$

35 The net shown is folded to make a cube. Which side is opposite Q?

U	T		
	S	R	
		Q	P

(A) R
(B) U
(C) S
(D) T

36 You are told that 219 x 582 + 171 = 127 629, answer the following question:
How many times can you subtract 582 from 127 629

(A) 171
(B) 582
(C) 219
(D) 291

37 If 75 men can erect a building in 12 days, how many men, working at the same rate, should erect the same building in 20 days?

(A) $3\frac{1}{5}$ (B) 18
(C) 45 (D) 25

38 Four more than twice a certain number is the same as 15 more than this same number. The number must be

(A) 8 (B) 7
(C) 9 (D) 11

39 The marked price of a desk was $480. Steven bought it at a discount of 20%. Chris bought the desk from Steven for 50% of the amount paid by Steven. How much did Chris pay for the desk?

(A) $384
(B) $192
(C) $96
(D) $150

40 There are 24 pupils in Primary 6A. The number of girls is twice that of boys. Each girl donates $4 and each boy donates $2 to a particular charity. How much money is collected altogether?

(A) $48
(B) $72
(C) $80
(D) $144

PAPER 10

1	0.2 h = ... min.	(A) 2 (B) 20 (C) 12 (D) 58
2	Which of the following is less than 50?	(A) $50 + \frac{1}{5}$ (B) $50 \times \frac{1}{5}$ (C) $50 \div \frac{1}{5}$ (D) $50 \times 1\frac{1}{5}$
3	What number is seventeen more than nine hundred and ninety?	(A) One thousand and seven (B) Nine hundred and twenty six (C) One thousand and sixteen (D) None of these
4	$5.44 \div 0.4 =$	(A) 1.36 (B) 13.6 (C) 136 (D) 2.176
5	The number of metres in $\frac{1}{8}$ km is	(A) 250 (B) 125 (C) 12.5 (D) 1250

6 - 7 In the figures given, the centres of the circles are shown.

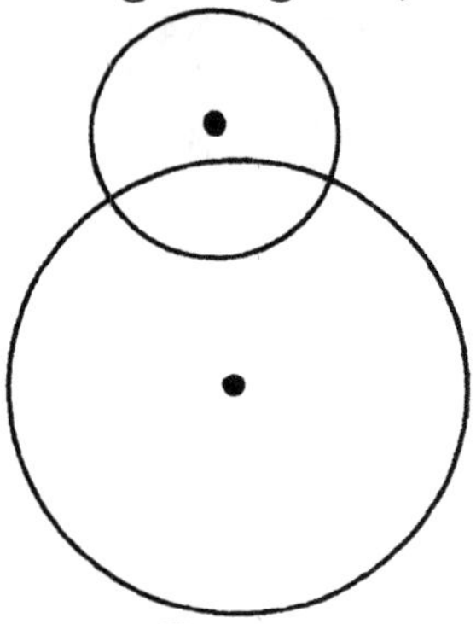

Figure 1

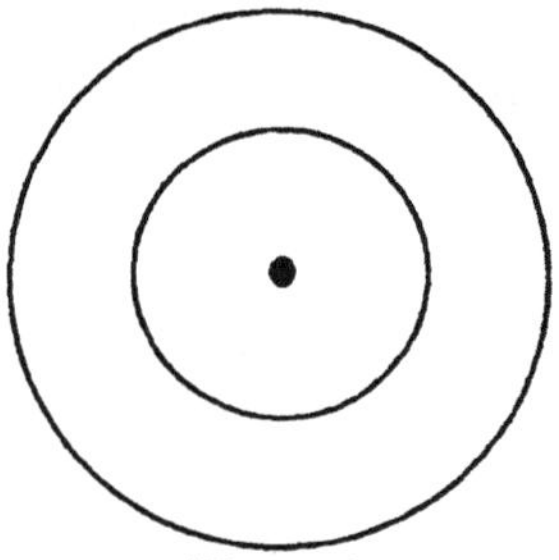

Figure 2

6	The number of axes of symmetry in figure 1 is	(A) 0 (B) 1 (C) 2 (D) 4
7	The number of axes of symmetry in figure 2 is	(A) 0 (B) 2 (C) 4 (D) infinite

8	The sum of 13.8m and 2m 5cm is	(A) 16.3m (B) 16m 3cm (C) 15m 13cm (D) 15.85m
9	The total weight of 15 identical balls is 11.25kg. What is the weight of each ball?	(A) 0.75kg (B) 168.75kg (C) 75g (D) 7.5kg
10	$\frac{1}{3}$ of a number exceeds $\frac{1}{4}$ of it by 5. What is that number?	(A) 30 (B) 45 (C) 60 (D) 120
11	0.5kg of flour costs 62.5 cents. What is the cost of 36kg of this flour?	(A) $37.25 (B) $40.50 (C) $45.50 (D) $45
12	Allan bought 12 twenty-cent stamps from the Post Office. He gave the cashier a ten dollar note. How much change should he receive?	(A) $8.40 (B) $7.40 (C) $8.60 (D) $7.60
13	The normal cost of 8 exercise books is $10.80. If Jane paid $40 for 4 dozen of these exercise books, how much discount did she receive on her purchase?	(A) $24.80 (B) $20 (C) $10.80 (D) $5
14	Mr. Henderson bought a T.V. set for $1850. He paid a deposit of $185 and the rest in equal 20 monthly instalments. How much did he pay each month?	(A) $92.50 (B) $85.50 (C) $83.25 (D) $60
15	A bookseller bought a book for $20 and sold it for $25. Express the gain as a percentage of the cost price.	(A) 125% (B) 20% (C) 25% (D) 80%
16	What percentage of the figure given is shaded?	(A) 75% (B) 50% (C) 35% (D) 25%

17	Which of the following is the greatest?	(A) $\frac{7}{8}$ (B) 86% (C) 0.85 (D) $\frac{4}{5}$
18	The front wheels of a cart have a circumference of 3 metres. The back wheels have a circumference of 4 metres. A white dot is painted on the rim of each wheel where it touches the ground. The cart is then driven 100 metres. Including the starting position, how many times will the white dots be on the ground together?	(A) 7 (B) 8 (C) 9 (D) 10
19	How many times must 357 be added to 419 to give 3632?	(A) 8 (B) 9 (C) 10 (D) 41
20	One number is 129 more than another. Their sum is 517. Their difference must be	(A) 517 (B) 129 (C) 323 (D) 388
21	A business earns $9471 in a year before expenses of $4182. The remainder is shared equally between three partners. Each partner would receive	(A) $3157 (B) $1394 (C) $1763 (D) $4551
22	The average cost of 9 eggs in one basket is 32 cents and the average cost of 8 eggs in another basket is 40 cents. What would be the average cost in cents of all the eggs?	(A) $5 + 32 \div 9$ (B) $72 \div 17$ (C) $608 \div 17$ (D) $9 \times 32 + 8 \times 40 \div 17$
23	What is the most likely number needed to complete the following series? 4, 7, 10, 13, ☐	(A) 17 (B) 14 (C) 16 (D) 15
24	Look at the following pattern: $T_1 = 1$ $T_2 = 1 + 3$ $T_3 = 1 + 3 + 5$ $T_4 = 1 + 3 + 5 + 7$ If this pattern continues, the most likely value for T_{10} is	(A) 10 (B) 100 (C) 50 (D) 60
25	When a certain number is doubled and has 13 added to the answer, the result is 37. The number is	(A) 16 (B) 14 (C) 12 (D) 10

26 - 27 In the British system of measurement, 3 feet = 1 yard, 1 mile = 1760 yards and 8 furlongs = 1 mile. 26 The number of feet in a mile is	(A) $586\frac{2}{3}$ (B) 5280 (C) 5180 (D) 2240
27 The number of feet in 5 furlongs is	(A) 3300 (B) 8448 (C) 1100 (D) None of these
28 - 29 A square prism has base area $9m^2$ and volume $36m^3$. 28 What is the perimeter of the base?	(A) 36m (B) 18m (C) 12m (D) 24m
29 What is the height of the prism?	(A) 324m (B) 25cm (C) $\frac{4}{9}$m (D) 4m
30 A bottle of fruit juice holds 375mL. A case holds 24 bottles. How many litres of fruit juice are there in a case?	(A) 8.9 (B) 9 (C) 2.25 (D) 8
31 How many pieces of plastic each of length $3\frac{1}{2}$ cm can be cut from a piece of length 1m 54cm?	(A) 44 (B) 22 (C) 66 (D) 88
32 A car completed a journey in 9h at 50km/h. What would have been the saving in time if the journey had been made at 60km/h instead?	(A) $7\frac{1}{2}$h (B) $\frac{1}{2}$h (C) $1\frac{1}{2}$h (D) $6\frac{2}{3}$h
33 If 1mL of water has mass 1 gram, what is the mass of 1 million litres of water?	(A) 1000t (B) 1000kg (C) 100kg (D) 10t
34 About 85% of a person's body mass is water. Katrina is a 12 year old girl of normal size. Which of these estimates could be the water mass in her body?	(A) 40g (B) 40kg (C) 4kg (D) 400kg

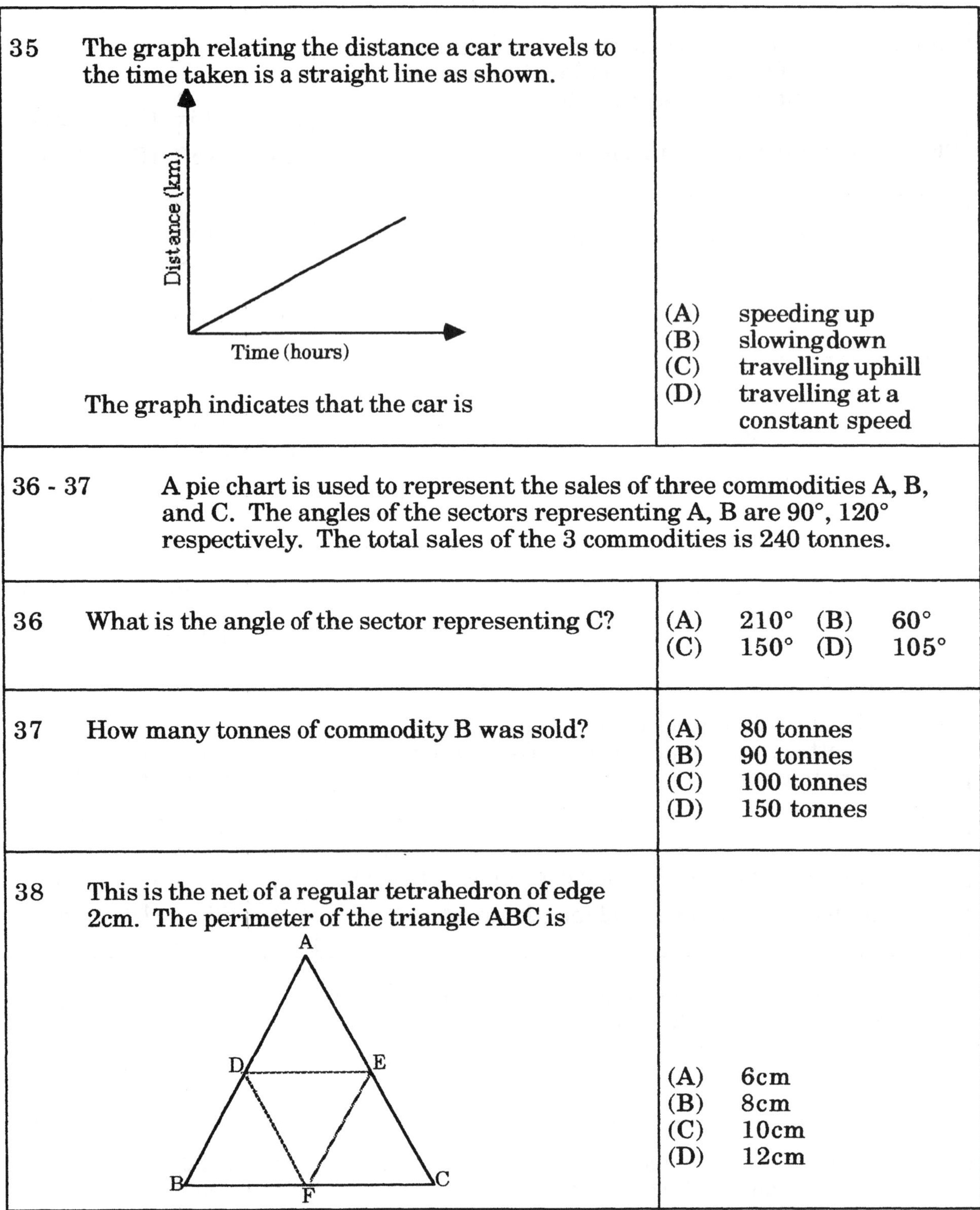

35 The graph relating the distance a car travels to the time taken is a straight line as shown.

The graph indicates that the car is

(A) speeding up
(B) slowing down
(C) travelling uphill
(D) travelling at a constant speed

36 - 37 A pie chart is used to represent the sales of three commodities A, B, and C. The angles of the sectors representing A, B are 90°, 120° respectively. The total sales of the 3 commodities is 240 tonnes.

36 What is the angle of the sector representing C?

(A) 210° (B) 60°
(C) 150° (D) 105°

37 How many tonnes of commodity B was sold?

(A) 80 tonnes
(B) 90 tonnes
(C) 100 tonnes
(D) 150 tonnes

38 This is the net of a regular tetrahedron of edge 2cm. The perimeter of the triangle ABC is

(A) 6cm
(B) 8cm
(C) 10cm
(D) 12cm

39 A builder wishes to construct a column of bricks 3m high to support part of a house. Each layer is to be like that shown in the sketch. 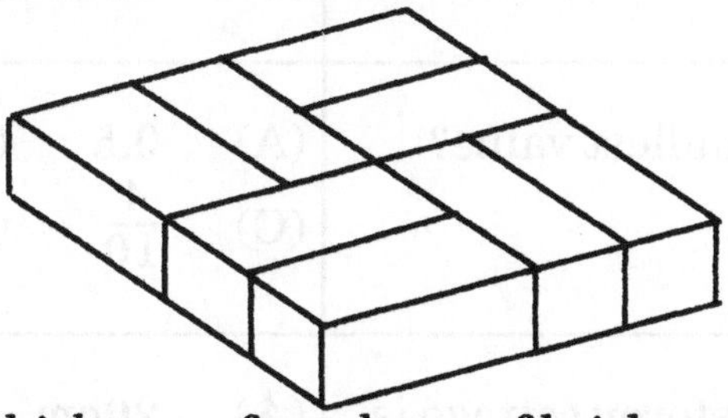 If the thickness of one layer of bricks and mortar is 10cm, how many bricks will be required?	(A) 80 (B) 160 (C) 240 (D) 480
40 A one-day cricket match is scheduled to have 8 hours playing time, four drink breaks each of 12 minutes, and a lunch break of 35 minutes. However, it finishes 29 minutes early. How long did the match last?	(A) 9h 23min (B) 9h 52min (C) 8h 54min (D) 8h 46min

FURTHER MATHEMATICS TESTS

PAPER 11

1	6.52L = ... mL	(A) 652 (B) 6052 (C) 6502 (D) 6520
2	Which of the following has the smallest value?	(A) 0.5 (B) 0.6 (C) $\frac{4}{10}$ (D) $\frac{4}{5}$
3	The height of an ordinary single storey cottage is approximately	(A) 20cm (B) 2m (C) 5m (D) 20m
4	The temperature has changed from 5°C to - 7°C. The temperature has	(A) increased by 2° (B) increased by 12° (C) decreased by 2° (D) decreased by 12°
5	Write down the sum of the following three numbers: four hundred and eleven, two thousand and thirteen and ten thousand and two	(A) 1626 (B) 12 426 (C) 3426 (D) 16 026
6	How many times the value of 2 is that of 5 in the number 5.32?	(A) 50 (B) 100 (C) 20 (D) 250
7	How many axes of symmetry does the parallelogram shown have?	(A) 0 (B) 1 (C) 2 (D) 4
8	A truck which can carry 1.5 tonnes is hired to remove 8 tonnes of soil from a building site. How many trips must the truck make?	(A) 5 (B) $5\frac{1}{3}$ (C) 6 (D) 6.5
9	For every 3 pens sold, a salesman received a commission of $10. How many pens must he sell in order to earn a commission of $240?	(A) 30 (B) 36 (C) 54 (D) 72
10	In a car park, there are 90 spaces for buses and 35 spaces for cars. What percentage of the spaces are for cars?	(A) 125% (B) 72% (C) 28% (D) $38\frac{8}{9}$%

11	The average of 40km, 28.5km, 25.9km, 13.6km and 35.7km is	(A) 28km 74m (B) 28km 740m (C) 28km 70m (D) 28km 40m
12	Royce had \$9.50. He spent \$2.40 on an exercise book and \$3.60 on a pen. What fraction of the original amount of money had he left?	(A) $\frac{7}{12}$ (B) $\frac{7}{19}$ (C) $\frac{13}{19}$ (D) $\frac{3}{19}$
13	Three identical containers hold 2.34L of liquid altogether. How much liquid will four such containers hold?	(A) 4.39L (B) 9.36L (C) 3.12L (D) 1.95L
14	A piece of ribbon was 2.72m long. Christina used $\frac{3}{8}$ of it to tie her hair. How many centimetres of ribbon were left?	(A) 170cm (B) 17cm (C) 1.7cm (D) 0.17cm
15	If 6 apples cost \$2.00, how many apples can I buy with \$15.00?	(A) 60 (B) 45 (C) 30 (D) 15
16	Story books are sold at a discount of 15%. How much does Stella have to pay for a story book which used to cost \$11.60?	(A) \$11.60 (B) \$9.86 (C) \$1.74 (D) \$6.80
17	$\frac{4}{5} - \frac{1}{2}$ as a decimal is	(A) 0.3 (B) 1.0 (C) 0.5 (D) 0.8
18	The digits of the number 4913 are arranged in descending order and then in ascending order. The difference between these two new numbers is	(A) 8082 (B) 4518 (C) 3564 (D) 8182
19	The sum of three consecutive odd numbers is 57. The largest of the three is	(A) 17 (B) 21 (C) 25 (D) 19
20	The pages of a book are numbered from 1 to 50. The number of times that the digit 4 will occur in the page numbers of the book will be	(A) 13 (B) 14 (C) 15 (D) 16

21	If the same one digit number greater than 2 is placed in each of the boxes, which of the following would be the largest?	(A) 2 x ☐ (B) ☐ x ☐ (C) ☐ x (☐ - 1) (D) ☐ x (2 x ☐ - 2)
22	The best approximation for 2.139 x 741.1 is	(A) 1400 (B) 15 (C) 1600 (D) 15000
23	Which statement correctly describes "Four more than twice the sum of 6 and 17 is decreased by 11"?	(A) 4 + 2 x 6 + 17 - 11 (B) 4 + 2 x (6 + 17 - 11) (C) (4 + 2 x 6) + 17 - 11 (D) 4 + 2 x (6 + 17) - 11
24	Town P is 480km from Town Q. A car goes from P to Q at 80km/h and returns at 48km/h. What is the average speed in km/h of the entire trip?	(A) 64 (B) 60 (C) 58 (D) None of these
25	What is the most likely number needed to complete the series? 7, 12, 19, 28, 39, ☐	(A) 50 (B) 52 (C) 53 (D) 54
26	A rectangle has perimeter of 6 metres. If the breadth is 1 metre, then the area is	(A) 5 m^2 (B) 3 m^2 (C) 2 m^2 (D) 1 m^2
27	The figure shows two squares of sides 3cm, 5cm. The shaded area is 2cm 3cm	(A) $16cm^2$ (B) $10cm^2$ (C) $6cm^2$ (D) $4cm^2$
28	The shape given is made up of	(A) a circle and a trapezium (B) a circle and a parallelogram (C) a semi-circle and a trapezium (D) a semi-circle and a parallelogram

29 - 31 A handyman has a 38m steel bar which he wants to cut and weld to make a framework in the shape of a cube (no fractions of a metre are required.)	
29 What is the side length of the largest cube that can be made?	(A) 6m (B) 4m (C) 3m (D) 2m
30 What length of the bar remains?	(A) 1m (B) 2m (C) 3m (D) 4m
31 The sides of this cube are enclosed. What volume of air does it contain?	(A) $8m^3$ (B) $27m^3$ (C) $9m^3$ (D) $64m^3$
32 A girl breathes 6.5 litres of air per minute. About how much air does she breathe in one day?	(A) 156L (B) 9360L (C) 390L (D) 23400L
33 A farmer uses 120L of water per minute from a tank to water a crop. After 10 minutes of watering the tank contains 15000L of water. What was the volume of water in the tank before he watered the crop.	(A) 14040L (B) 14880L (C) 15120L (D) 16200L
34 - 37 One hundred and twenty five blocks each a 1cm cube, are placed on a table, so as to form a solid cube.	
34 State the length of the edge of this cube.	(A) 4cm (B) 5cm (C) 6cm (D) 7cm
35 The top and the four side faces of this large cube are now painted red. How many of the original 1cm cubes have just one face painted red?	(A) 60 (B) 45 (C) 57 (D) 64
36 How many more 1cm cubes would need to be added to the 125 cubes to form the next larger solid cube?	(A) 101 (B) 91 (C) 11 (D) 216
37 If the outer layer of cubes is peeled off the original solid cube, to form another solid cube, what is the total volume of the cubes peeled off.	(A) $61cm^3$ (B) $100cm^3$ (C) $56cm^3$ (D) $98cm^3$

38 Which of the following nets could be folded to make a square pyramid?

(A) (B) (C) (D)

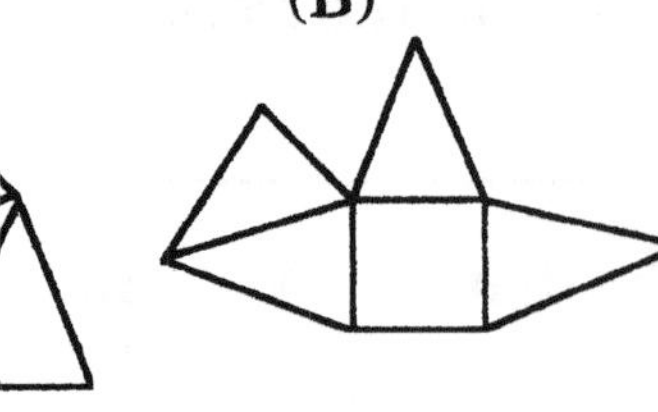

39 The figure shown consists of squares each of side 3cm. It is cut out and folded along the broken lines to form a "hollow" solid. What is the sum of the internal and external surface areas of this solid?

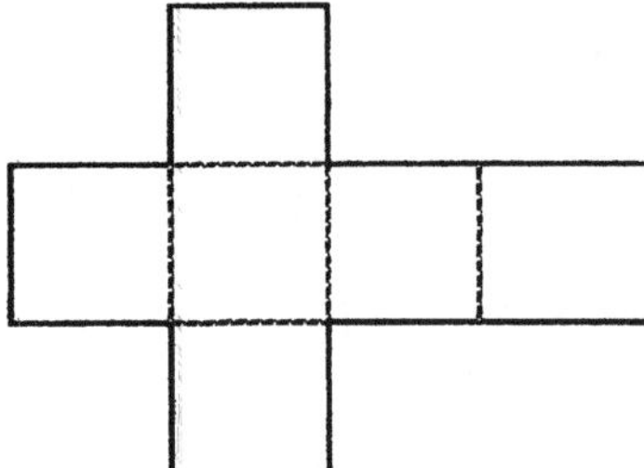

(A) 27cm^3
(B) 54cm^2
(C) 108cm^2
(D) 90cm^2

40 The value of $1\frac{4}{5} + 3\frac{2}{3} + 6\frac{11}{15}$ is

(A) $12\frac{1}{5}$ (B) $11\frac{1}{5}$
(C) $10\frac{1}{5}$ (D) $12\frac{2}{15}$

PAPER 12

1	Which of the following is different in value from the rest? $\frac{6}{12}, \frac{2}{4}, \frac{5}{10}, \frac{8}{15}$	(A) $\frac{8}{15}$ (B) $\frac{5}{10}$ (C) $\frac{2}{4}$ (D) $\frac{6}{12}$
2	Which of the following is not equal to 25%?	(A) 25 out of 100 (B) 0.25 (C) $\frac{1}{4}$ (D) 25
3	$3 \times 10 + 5 + \frac{2}{10} + \frac{17}{100} =$	(A) 35.37 (B) 35.217 (C) 3.527 (D) 350.37
4	$\frac{7}{20} = \frac{\square}{100} = \square$%. What are the missing numbers in the boxes?	(A) 7; 7 (B) 35; 35 (C) 14; 14 (D) 20; 20
5	Two numbers multiply to give 84 and add to give 19. The smaller of the two numbers will be	(A) 1 (B) 12 (C) 19 (D) 7
6	20% of 3 metres is	(A) 3cm (B) 6cm (C) 60cm (D) 30cm
7	A Babylonian tablet is dated by archaeologists as 1780 B.C. What is the approximate age of the tablet?	(A) 1000 years (B) 2000 years (C) 3000 years (D) 4000 years
8 - 9	Given that $19^2 = 361$ and $19^3 = 6859$, then for a cube of side 0.19m, the	
8	area of each face is	(A) $0.361m^2$ (B) $3.61m^2$ (C) $0.0361m^2$ (D) $0.00361m^2$
9	volume of the cube is	(A) $68.59m^3$ (B) $0.6859m^3$ (C) $0.06859m^3$ (D) $0.006859m^3$

10 Three children paid a total of $15.90 to see a movie. How much would 20 children pay to see the same movie?	(A) $106 (B) $318 (C) $118 (D) $53
11 - 13 The travel graph given represents the journey of a car. 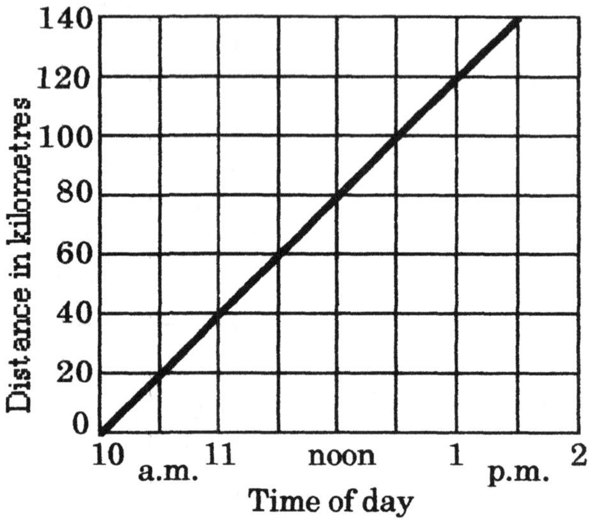 Distance in kilometres: 0, 20, 40, 60, 80, 100, 120, 140 Time of day: 10 a.m., 11, noon, 1 p.m., 2	
11 How far was the car from its starting point at 11:15a.m.?	(A) 50km (B) 60km (C) 45km (D) 55km
12 At what time was the car 110 km from its starting point?	(A) 1p.m. (B) 12:15p.m. (C) 12:30p.m. (D) 12:45p.m.
13 What was the average speed of the car?	(A) 30km/h (B) 40km/h (C) 50km/h (D) 60km/h
14 The number of seconds in 1 day is	(A) 1440 (B) 3600 (C) 86 400 (D) 31 536 000
15 If the first and last digits of the number 2196 are interchanged, then the new number is	(A) larger than the old number by 3996 (B) three times as large as the old number (C) smaller than the old number by 3996 (D) none of these

16	The best approximation for 1.37 x 546.1 is	(A) 7 (B) 70 (C) 700 (D) 7000
17	What percentage of the set of shapes are circles?	(A) 10% (B) 20% (C) 30% (D) 40%
18	A motorist drives for 45 minutes at 72km/h. The distance travelled in kilometres would be	(A) 54 (B) $1\frac{3}{5}$ (C) 96 (D) None of these
19	In sharing \$180 with Steven, Don gets one third. How much must Steven give Don in order that they will have the same amount?	(A) \$120 (B) \$80 (C) \$60 (D) \$30
20	Complete the series of numbers below. 225, 45, 9, ... , 0.36	(A) 4.5 (B) 3 (C) 1.8 (D) 0.9
21	What fraction of the whole figure is shaded?	(A) $\frac{1}{5}$ (B) $\frac{1}{6}$ (C) $\frac{2}{9}$ (D) $\frac{1}{4}$
22	Andrew is 120cm tall. Ian is $1\frac{1}{3}$ times as tall as Andrew. What is the sum of their heights?	(A) 280cm (B) 180cm (C) 240cm (D) 160cm
23	In $\frac{24 + 4 \times 4}{8}$ ❑ $\frac{3 \times (9 + 2)}{10}$, the missing sign is	(A) = (B) < (C) > (D) None of these

24	Simon's wage is \$750. He spends \$500 and saves the rest. What fraction of the income does he save?	(A) $\frac{1}{5}$ (B) $\frac{1}{3}$ (C) $\frac{1}{2}$ (D) $\frac{1}{4}$
25	The first multiple of 7 greater than 995 is	(A) 1001 (B) 997 (C) 1005 (D) 999
26	Three consecutive numbers add to 147. The smallest of the three will be	(A) 48 (B) 49 (C) $73\frac{1}{2}$ (D) 50
27	Which numerical statement correctly describes the statement "Three times the sum of 5 and the product of 2 and 7 is decreased by 1"?	(A) 3 x (5 + 2 x 7) - 1 (B) 3 x 5 + 2 x 7 - 1 (C) 3 x (5 + 2) x 7 - 1 (D) 3 x (5 + 2 x 7 - 1)
28	The figures given consist of equal squares. The figures have 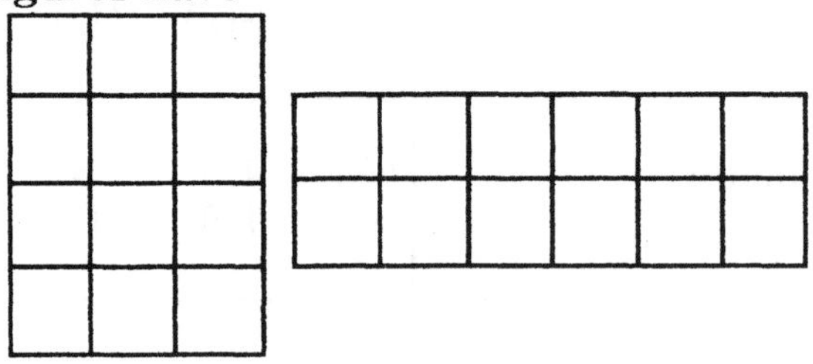	(A) different areas and the same perimeter (B) different areas and different perimeters (C) the same area and the same perimeter (D) the same area and different perimeters
29	The obtuse angle between the directions North and South East is	(A) 315° (B) 225° (C) 145° (D) 135°
30	There are 11.24L of chemical stored in a container. If 560mL are removed and later 390mL are put back, what is the quantity of chemical in the container now?	(A) 12.94L (B) 11.07L (C) 9.54L (D) 11.41L
31	How many cm^2 are there in $24m^2$?	(A) 0.0024 (B) 0.24 (C) 2400 (D) 240 000
32	A rectangular paddock is 2km x 1km. The area is	(A) 2ha (B) 20ha (C) 200ha (D) 1000ha

33 A ship covers 52 nautical miles in 4h 20min. What is its speed in knots? [1 knot = 1 nautical mile per hour]	(A) 10 (B) 11 (C) 12 (D) 13
34 A substance weighs 5kg per litre. How many grams per cm^3 is this?	(A) 5 (B) 50 (C) 500 (D) 0.5
35 - 36 Jenny's salary is spent as shown in the graph. Rent 25% Extras 5% Savings 60% Living Expenses 35 What fraction is her Living Expenses of her Savings?	(A) $\frac{1}{10}$ (B) $\frac{1}{6}$ (C) $\frac{3}{5}$ (D) $\frac{6}{1}$
36 If her weekly Rent is \$150, what are Jenny's weekly Living Expenses?	(A) \$360 (B) \$600 (C) \$300 (D) \$720
37 - 38 A square pyramid is made up of identical solid spheres such that each sphere touches it neighbours. Given that the bottom layer contains 49 spheres, then 37 the number of layers in the pyramid is	(A) 5 (B) 6 (C) 7 (D) 8
38 the number of spheres in the pyramid is	(A) 91 (B) 140 (C) 204 (D) 139
39 There are 3 toy cars. The first toy car costs \$6.90. The second toy car costs twice as much as the first. The average cost of the three toy cars is \$15.80. How much is the third toy car?	(A) \$13.80 (B) \$22.70 (C) \$25 (D) \$26.70
40 Grace spent 0.5 of her money on comics and half of remainder on some exercise books. If she had \$12.50 left, how much had she at first?	(A) \$45 (B) \$40 (C) \$35 (D) \$50

FURTHER MATHEMATICS TESTS

PAPER 13

1	Round off \$48 550 to the nearest thousand dollars.	(A) \$47 000 (B) \$48 000 (C) \$49 000 (D) \$50 000
2	2.02 x 400 =	(A) 808 (B) 880 (C) 8800 (D) 80.8
3	Decrease 4m 50cm by 10%.	(A) 4m 50cm (B) 4m 45cm (C) 4m 5cm (D) 4m 55cm
4	$2\frac{1}{4} \div 0.5 =$	(A) $1\frac{1}{8}$ (B) $\frac{9}{20}$ (C) $4\frac{1}{2}$ (D) $\frac{2}{9}$
5	Which of the following diagrams could represent the front view of the solid shown in the diagram? (A) (B) (C) (D)	
6	A roll of aluminium foil is 30cm wide and 16m long. The area of the roll is	(A) $4.8m^2$ (B) $48m^2$ (C) $480m^2$ (D) $4800m^2$
7	Three girls weigh a total of 94.8kg. If two of them weigh 27.6kg and 35.8kg respectively, what is the weight of the third girl?	(A) 31kg (B) 31.5kg (C) 31.4kg (D) 31.8kg
8	I buy $1\frac{1}{2}$ dozen pens. If each pen costs 75c, what change will I receive from a \$20 note?	(A) \$13.50 (B) \$15 (C) \$7.50 (D) \$6.50

9	A square of side length 1m is divided up into 20 equal parts and 15 of these parts are shaded in. The area shaded as a decimal fraction of a square metre is	(A) 0.35m^2 (B) 0.60m^2 (C) 0.75m^2 (D) 0.55m^2
10	Susan saved \$45.50 in 5 months. At this rate, how much could she save in a year?	(A) \$227.50 (B) \$145.50 (C) \$109.20 (D) \$91.50
11	A book-mark costs \$0.60. Sam bought 3 of them, Huck bought 2 more than Sam and Jim bought 5 more than Huck. How much did they spend altogether?	(A) \$3.00 (B) \$4.20 (C) \$6.00 (D) \$10.80
12	If 75% of a number is 18, what is the number?	(A) 36 (B) 30 (C) 25 (D) 24
13	What same number would you put in each box to make the following number sentence true? 4 x (□ + 5) = 3 x (□ + 8)	(A) 3 (B) 4 (C) 8 (D) 2
14	The number 153 has the strange property that $153 = 1^3 + 5^3 + 3^3$. Which of the following numbers has the same property?	(A) 154 (B) 253 (C) 136 (D) 407
15	I have 15 banknotes. Each of them is either a \$10 or a \$5 note. If their total value is \$110, the number of \$10 notes is	(A) 6 (B) 8 (C) 7 (D) 9
16	If 8 lemons cost \$1.28, then 15 lemons would cost	(A) \$2.40 (B) \$0.68 (C) \$8.96 (D) \$2.56
17	A girl can run 7 kilometres in 56 minutes. If she ran at the same rate, how long would she take to run 8 kilometres?	(A) 49 minutes (B) 52 minutes (C) 1hour 4 minutes (D) 1hour 20 minutes
18	A strange signal is received from outer space: MJKLMJKLMJKL.... The letters M, J, K, L are repeated over and over. If this pattern were to continue, what would the 5218th letter be?	(A) M (B) J (C) K (D) L

19 I made a necklace from red, blue, yellow and green beads. I started with a red bead, then blue, then yellow and then green and repeated the pattern until I ran out of beads. The last bead was red. What can be said about the number of beads on the necklace?

(A) It was even
(B) If you divided the number by 4, the remainder would be 3
(C) The number would be prime
(D) The number was 1 more than a multiple of 4

20 What is the most likely number needed to complete the following pattern?

2	5	19	14	8	13
18	15	1	6	12	

(A) 14
(B) 7
(C) 9
(D) None of these

21 The best approximation for $\frac{4.13 \times 91.65}{74.1 + 28.5}$ is

(A) 4 (B) 37
(C) 401 (D) 2

22 18 km/h is equivalent to

(A) 18m/s (B) 30m/s
(C) 5m/s (D) 6m/s

23 - 24 The area of a square is 400 cm^2.

23 The side of the square is

(A) 2m (B) 20cm
(C) 50cm (D) 24cm

24 The area of the square is increased by one fifth. The side of this new square is nearest to

(A) 21cm (B) 22cm
(C) 23cm (D) 24cm

25 A square sheet of cardboard is cut vertically into two equal pieces, each of perimeter 24cm. What is the perimeter of the original sheet?

(A) 48cm (B) 24cm
(C) 32cm (D) 36cm

26 A container containing 6 litres of water is used to fill 24 bottles of equal size. How much water does each bottle hold?

(A) 144mL (B) 25mL
(C) 250mL (D) 375mL

27 The petrol tank of a car holds 64 litres when full. Before a trip the petrol gauge shows $\frac{3}{4}$ full and after the trip it shows $\frac{1}{4}$ full. How many litres of petrol were used on the trip?

(A) 32L
(B) 128L
(C) 16L
(D) 48L

28 How many cubes each of side 3cm are required to fill a rectangular box of dimensions 12cm x 9cm x 6cm?	(A) 12 (B) 6 (C) 24 (D) 48
29 - 30 A large rectangular tank of base $3\frac{1}{2}$ m x $2\frac{1}{2}$ m contains water to a height of 4m. (Note: $1m^3$ of water equals 1000 litres, and 1 mL of water weighs 1 gram.)	
29 The capacity of water in the tank is	(A) 35L (B) 3500L (C) 35 000L (D) 25 000L
30 As the result of a downpour, 400 litres of water runs off in the tank, what is the weight of this run off?	(A) 4kg (B) 40kg (C) 400kg (D) 4000kg
31 Which of the following transforms the triangle PQR on to the triangle P Q'R' ? Q Q' P R R'	(A) An anti clockwise rotation of 90° about P (B) A clockwise rotation of 90° about P (C) A reflection in the line bisecting the angle QPQ' (D) A rotation through 180° about the midpoint of the line segment QQ'
32 Which of the following transformations of the figure Z shown, will result in an image identical with the original figure? L O m	(A) a rotation of 180° about O (B) a rotation of 90° about O (C) a reflection in the line m (D) a reflection in the line L

33 A triangle and its image are always the same size and shape under which of the following transformations? I - Translation II - Rotation III - Reflection IV - Enlargement	(A) I, II and III (B) I, III and IV (C) I, II, III and IV (D) II, III and IV
34 - 35 A girl builds a series of block towers as shown; each block is a cube of side 2cm. 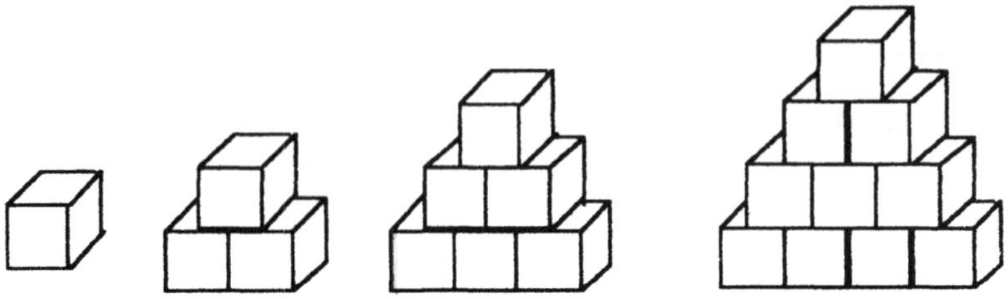 34 To change a seven storey tower to a ten storey tower, how many extra blocks are needed?	(A) 30 (B) 27 (C) 24 (D) 17
35 What is the total volume of an eight storey tower?	(A) $36cm^3$ (B) $14cm^3$ (C) $288cm^3$ (D) $216cm^3$
36 - 37 The graph shows the distribution of marks gained in a test by a group of students. 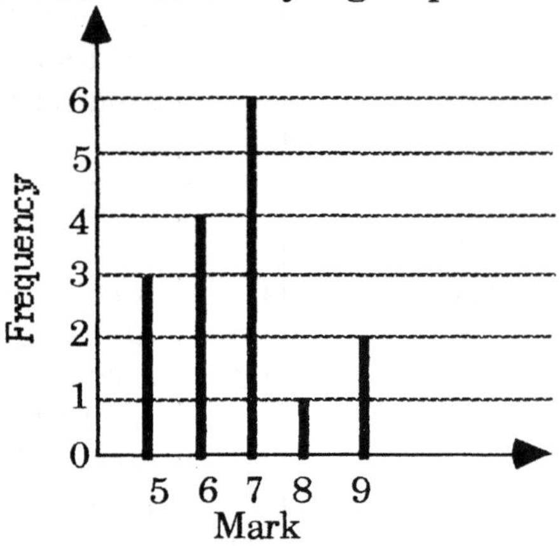 Frequency: 0 1 2 3 4 5 6 Mark: 5 6 7 8 9 36 How many students are there?	(A) 5 (B) 16 (C) 21 (D) 35
37 The total number of marks achieved by the group in the test is	(A) 35 (B) 21 (C) 107 (D) 127

38	A bag of rice weighs 8kg. The rice is repacked into smaller packets each weighing 0.6kg. How much rice will remain after the greatest possible number of packets have been filled?	(A) 2kg (B) 0.2kg (C) 4kg (D) 0.4kg
39	A motorist drives for 75 minutes at 60km/h and then drives at 90km/h for another 15 minutes. The average speed for the entire journey in km/h would be	(A) 150 (B) 65 (C) 75 (D) None of these
40	The sum of the ages of two sisters is 56 years and the difference in their ages is $\frac{1}{7}$ of the sum. Find the age of the older sister in years.	(A) 48 years (B) 36 years (C) 40 years (D) 32 years

FURTHER MATHEMATICS TESTS

PAPER 14

1	$\frac{4}{5} = \frac{32}{\square}$. The correct number in the box is	(A) 20 (B) 30 (C) 35 (D) 40
2	Which of the following is an improper fraction?	(A) $\frac{1}{4}$ (B) $\frac{4}{5}$ (C) $\frac{11}{10}$ (D) $\frac{2}{3}$
3	Express $\frac{3}{8}$ as a decimal.	(A) 0.125 (B) 0.375 (C) 0.575 (D) 0.825
4	Which of the following is the correct way of rounding off 4193 to the nearest hundred?	(A) 4190 (B) 4290 (C) 4200 (D) 4100
5	The length of an airliner used to transport passengers would be about	(A) 5m (B) 50m (C) 500m (D) 5km
6	Which solid shown is a prism? (A) (B) (C) (D)	
7	How many axes of symmetry has an equilateral triangle?	(A) 8 (B) 1 (C) 2 (D) 3
8	Mrs Johnson filled four 1.15L bottles with oil and had 264mL of oil left. How much oil had she at first?	(A) 4.864L (B) 4.264L (C) 4.6L (D) 0.264L
9	A woman did a piece of work during February 1994. Assuming she did an equal amount each day, what fraction of it remained unfinished after 21 days?	(A) $\frac{3}{4}$ (B) $\frac{1}{4}$ (C) $\frac{21}{29}$ (D) $\frac{8}{29}$
10	Mrs. Miller saved a total of $120 in the first 5 months of a year. She saved $66 in the sixth month. On the average, how much did she save per month over the period?	(A) $30 (B) $31 (C) $33 (D) $34

11	Fifty pupils and two teachers went to the museum. Each pupil paid $1.80 for admission. Each teacher paid $3.60. How much did they pay altogether?	(A) $90 (B) $95 (C) $97 (D) $97.20
12	There are 24 girls and 18 boys in a group. Find the percentage of girls in the group.	(A) $57\frac{1}{7}$% (B) 42% (C) 58% (D) 55%
13	Which of the following is the greatest?	(A) 100 ÷ 7 (B) 0.1 x 0.7 (C) 100% of 0.7 (D) 0.7 x 10
14	Janet bought a second hand car for $6000. After using it for two years, she sold it at a profit of 10% of the cost price. At what price did she sell?	(A) $6010 (B) $6060 (C) $6006 (D) $6600
15	Seven years ago a man was twice as old as his sister. If his sister is now 15 years old, how old, in years, is the man?	(A) 15 (B) 17 (C) 19 (D) 23
16	The distance between the consecutive whole numbers on a number line is 1 centimetre. What is the least number of whole numbers which could lie in a 10 centimetre interval?	(A) 9 (B) 10 (C) 11 (D) 8
17	What number should be placed in the box to make the following statement true? ☐ ÷ 8 x 5 = 10	(A) 8 (B) 2 (C) 12 (D) 16
18 - 20	A teacher painted the letters of the alphabet, in alphabetical order, on the concrete of the school playground. The letters were arranged so that if you stood on the letter A and walked along the line of letters, you would step on a new letter with each step you took.	
18	Berrima stood on the letter D, took 10 steps to the right (i.e. towards the letter Z) then 6 steps to the left and then 14 steps to the right. On what letter was she now standing?	(A) S (B) D (C) R (D) V

<table>
<tr><td>19 Bruce stood on the letter X and Kel stood on the letter F. At the same time and at the same rate they started walking towards each other. On what letter would they meet?</td><td>(A) L
(B) O
(C) N
(D) They won't meet on a letter</td></tr>
<tr><td>20 Karyn stands on the A and Sela stands on the Z. At the same time they walk towards the other end. If Karyn walked twice as fast as Sela, what letter would Sela be stepping on when Karyn was stepping on the K?</td><td>(A) V
(B) T
(C) U
(D) E</td></tr>
<tr><td>21 A "palindromic" number is one which reads the same forwards as backwards. For example 32123 is palindromic. If 32123 is multiplied by certain single digit numbers, the answers are also palindromic. Those numbers are</td><td>(A) 2 only
(B) 3 only
(C) 2 and 3 only
(D) 1 and 2 and 3 only</td></tr>
<tr><td>22 What is the most likely number needed to complete the following pattern?
<table><tr><td>8</td><td>3</td><td>10</td><td>5</td><td>7</td><td>1</td></tr><tr><td>8</td><td>18</td><td>4</td><td>14</td><td>10</td><td></td></tr></table></td><td>(A) 16
(B) 18
(C) 22
(D) None of these</td></tr>
<tr><td>23 I have 13 coins in my pocket. They are all either 20cent or 50cent coins. Their value is $5.30. The number of 20cent coins must be</td><td>(A) 9 (B) 6
(C) 8 (D) 4</td></tr>
<tr><td>24 If 9 shirts cost $240, the number of shirts which could be bought for $560 is</td><td>(A) 3 (B) 21
(C) 18 (D) 24</td></tr>
<tr><td>25 - 26 A group of people is selected so that the names of four of them start with the letter A, four of them start with the letter B and so on through the alphabet.

25 If they were to stand in alphabetical order, with what letter would the name of the 87th person start?</td><td>(A) T
(B) U
(C) V
(D) W</td></tr>
<tr><td>26 If we started to count from the last person, i.e. we counted in reverse alphabetical order, with what letter would the 70th person's name begin?</td><td>(A) I
(B) J
(C) H
(D) None of these</td></tr>
</table>

27 A piece of wire 26cm long is bent to form an isosceles triangle with base 8cm. The length of each equal side is	(A) 5cm (B) 10cm (C) 9cm (D) 18cm
28 The figure given is made up of an equilateral triangle and a rectangle. What is its perimeter? 11cm 9cm	(A) 40cm (B) 49cm (C) 51cm (D) 58cm
29 On a scale drawing, 2cm represents 360 metres. The scale used is	(A) 1 : 180 (B) 1 : 1 800 (C) 1 : 18 000 (D) 1 : 180 000
30 - 31 A square prism, open at the top, has base of side 6cm and height 10cm. 30 Its volume is	(A) $60cm^3$ (B) $360cm^3$ (C) $600cm^3$ (D) $120cm^3$
31 Its external surface area is	(A) $252cm^2$ (B) $240cm^2$ (C) $276cm^2$ (D) $312cm^2$
32 - 33 Rodriguiz takes 40 paces to cover a distance of 32 metres. 32 What is the average length of his pace?	(A) 8cm (B) 80cm (C) 1.28m (D) 128cm
33 He takes 90 paces along the side of a building. What is the approximate length of the building?	(A) 7.2m (B) $14\frac{2}{9}$ m (C) 720m (D) 72m
34 The reflex angle between the direction NW and S is	(A) 135° (B) 225° (C) 270° (D) 245°

35 - 36 The graph shown represents the travel graph of a return bus trip from town P to town Q.

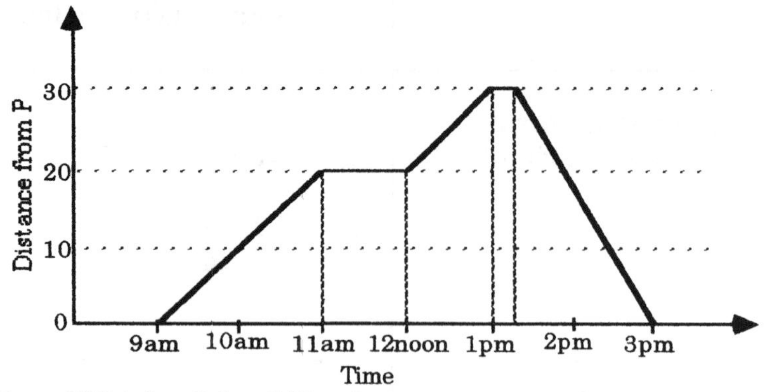

35 Which of the following statements is not a true interpretation of the graph.

(A) The return trip was alldownhill.
(B) The bus stopped in town Q for less that 30 minutes.
(C) The bus was not moving between 11:00a.m. and 12noon.
(D) The average speed of the return trip to town P was faster than the average speed of the trip to town Q.

36 The average speed of the bus for the whole trip is

(A) 5km/h (B) $7\frac{1}{2}$ km/h
(C) 10km/h (D) $12\frac{1}{2}$km/h

37 Which shapes given below could be cut out and folded along the dotted lines to form a triangular prism?

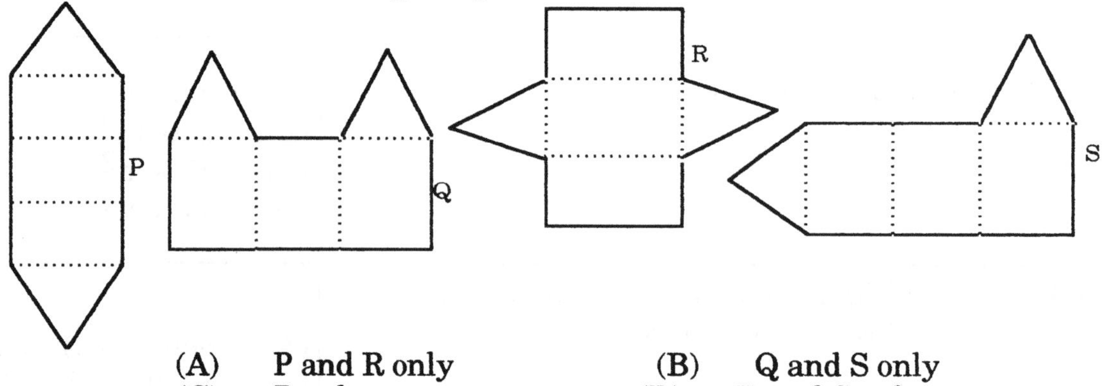

(A) P and R only (B) Q and S only
(C) R only (D) R and S only

38 The diagram shows a counter located at position (W,p) on a board. The counter slides 5 squares along the diagonal.

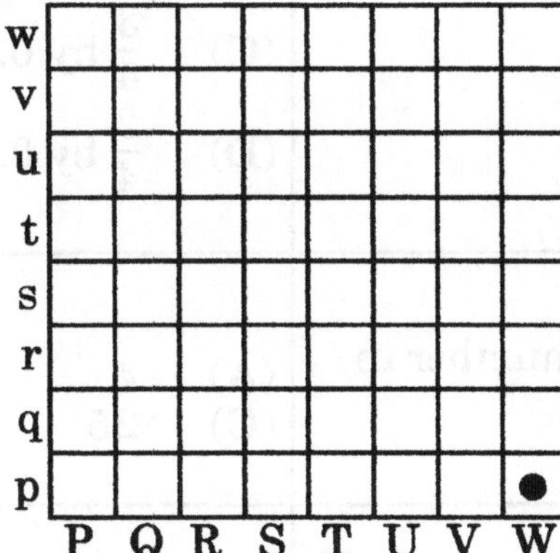

Its new position is

(A) (R,p)
(B) (S,t)
(C) (R,u)
(D) (U,r)

39 The pie chart shows how income from the sale of a compact disc (CD) is shared.

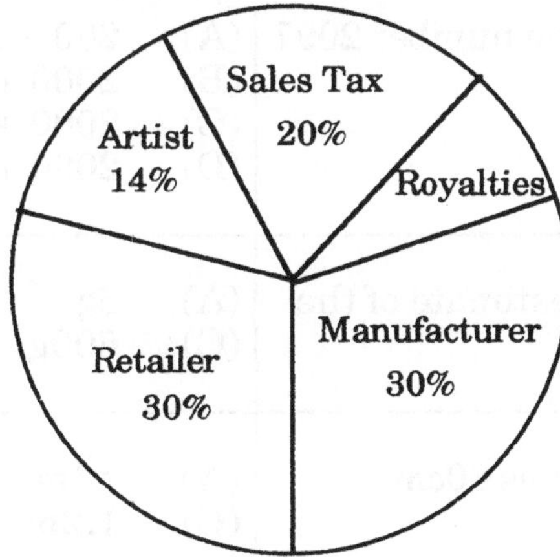

If the CD sells for $15, how much of this goes to Royalties?

(A) $9
(B) 90c
(C) $2.50
(D) 40c

40 In the sketch given, how many more cartons are needed to fill the trolley?

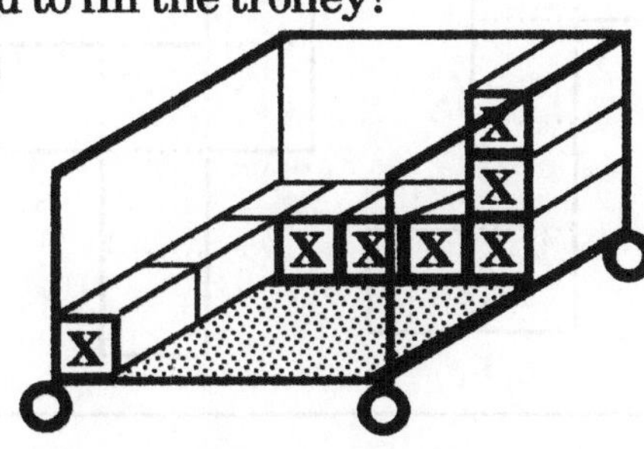

(A) 45
(B) 36
(C) 33
(D) 28

PAPER 15

1	Which is larger and by how much, 0.6 or $\frac{3}{4}$?	(A) 0.6 by 0.05 (B) 0.6 by 0.15 (C) $\frac{3}{4}$ by 0.05 (D) $\frac{3}{4}$ by 0.15
2	$4.25 = 4 + \frac{1}{\square}$. What is the missing number in the box?	(A) 4 (B) 10 (C) 25 (D) 100
3	If the sun rises at 5:51a.m. and sets at 7:09p.m, then there was daylight for	(A) 13 hours 18 minutes (B) 12 hours 9 minutes (C) 13 hours (D) None of these
4	When written in expanded form, the number 2097 becomes	(A) 200 + 90 + 7 (B) 2000 + 900 + 7 (C) 2000 + 90 + 7 (D) 2000 + 900 + 70
5	Which of the following is the best estimate of the weight of a hen's egg?	(A) 6g (B) 60g (C) 600g (D) 6000g
6	Which length is half as large again as 80cm?	(A) 40m (B) 60cm (C) 1.2m (D) 1.6m

7 When cut out and folded along the broken lines, which shape given will not form a cube?

(A) (B) (C) (D)

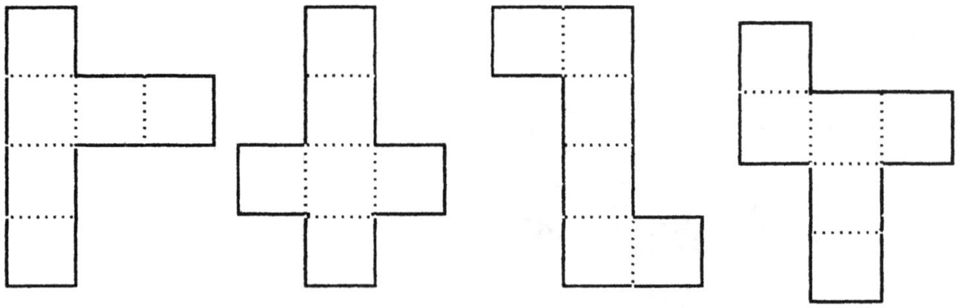

8 How many axes of symmetry does this figure have?	(A) 0 (B) 1 (C) 2 (D) 4
9 Which of these solids is a prism? (A) (B) (C) (D)	
10 Two circles have the same centre. The larger circle is of diameter 3.0cm and the distance between the circles is 0.8cm. What is the radius of the smaller circle?	(A) 2.3cm (B) 1.1cm (C) 2.2cm (D) 0.7cm
11 Which one of the following sets of decimals is arranged from the smallest to the largest?	(A) 0.07, 0.5, 0.14, 0.8 (B) 0.5, 0.07, 0.14, 0.8 (C) 0.14, 0.5, 0.07, 0.8 (D) 0.07, 0.14, 0.5, 0.8
12 A girl walked at the rate of 4km/h for $1\frac{1}{2}$h. Then she cycled at the speed of 10km/h for $\frac{1}{2}$h. What is her average speed for this trip?	(A) $4\frac{1}{2}$km/h (B) 11km/h (C) $5\frac{1}{2}$km/h (D) 9km/h
13 Three fifths of the workers in a factory are men and the rest are women. If there are 180 women, how many are men?	(A) 450 (B) 90 (C) 360 (D) 270
14 Margaret had $680. She spent 15% of her money on food. She gave 25% of the remainder to her two sons and saved the rest. How much did she save?	(A) $578 (B) $216.75 (C) $433.50 (D) $170

15	A school ground is 240m long and 80m wide. Find the cost to fence it at $10.50 per metre.	(A) $201 600 (B) $3360 (C) $2520 (D) $6720
16	Betty paid $15 for the box of 60 oranges. 10% of the oranges were rotten. She sold the remaining oranges at 40 cents each. Find the profit.	(A) $21.60 (B) $6.60 (C) $24 (D) $12
17	20% of the water in a full container with a capacity of 200 cm^3 is required to fill a jug containing 55cm^3 of water. What is the capacity of the jug?	(A) 55mL (B) 75cm^3 (C) 95mL (D) 110cm^3
18	How many digits would be needed to number the pages of a book with 256 pages, starting with page 1?	(A) 256 (B) 570 (C) 495 (D) 660
19	If 15% of a sum of money is $96, what is 25% of the sum of money?	(A) $640 (B) $160 (C) $24 (D) $14.40
20	A video recorder costs $1280. In how many month's time will Edward save enough to buy it if he saves 10% of his monthly salary of $1 600?	(A) 2 (B) 4 (C) 8 (D) 12
21	Two girls mark out a race track in the shape of an equilateral triangle. The girls start the race at P, the first girl running clockwise round the triangle and the second running anticlockwise round the triangle. If the first girl runs twice as fast as the second girl, where will they meet? R P Q	(A) At R (B) At Q (C) Between R and Q (D) You need more information
22	A car travels 840km in 6 days. How far, in kilometres, will it travel in 7 days at the same rate?	(A) 720 (B) 980 (C) 960 (D) 1020

23	A Large Number is divisible by both 3 and 4. The next smallest number that the Large Number is certainly divisible by is	(A) 5 (B) 6 (C) 12 (D) It is impossible to tell
24	Four people in a room shake hands with each other once. How many handshakes were there?	(A) 2 (B) 4 (C) 5 (D) None of these
25	Each of the following numbers uses the digits 1, 2 and 3. Which is the largest?	(A) 321 (B) 12^3 (C) 13^2 (D) 2^{13}

26 The two 4's in the numbers 14 253 and 93 842 have different values. Which of the following is true?
(A) The 4 in the first number is 1000 times as large as the 4 in the second
(B) The 4's differ in value by 396
(C) The 4 in the first number measures the number of 10 000's and the 4 in the second measures the number of 10's.
(D) None of these

27	The distance between consecutive telegraph poles along a road is 50 metres. What is the greatest number of telegraph poles which could lie in 1 kilometre of road?	(A) 19 (B) 20 (C) 21 (D) 22
28	What same number should be placed in each box to make the following statement true? $\square \times 5 + 9 \times \square = 28$	(A) 1 (B) $\frac{1}{2}$ (C) 4 (D) 2
29	A mail sorter sorts 6480 letters in 4h 30min. How many letters per minute is this?	(A) 18 (B) 24 (C) 36 (D) 48
30	A certain liquid weighs 800 grams per litre. How many cm^3 of the liquid weigh 1 kilogram?	(A) 750 (B) 1000 (C) 1250 (D) 1500
31	A square paddock has area 1 ha. What is the length of each side?	(A) 100m (B) 1000m (C) 10m (D) 32m
32	If 90% of a certain amount q of liquid is 360mL, what is the value of q?	(A) 40mL (B) 4L (C) 32.4L (D) 400mL

33	A closed cube has surface area 96cm^2. What is the length of the side of the cube?	(A) 2cm (B) 3cm (C) 4cm (D) 6cm
34	I sell goods costing me 1c per g for $12 per kilogram.	(A) I lose $2.00 per kg (B) I gain $2.00 per kg (C) I neither gain nor lose (D) I gain $11.00 per kg
35	A stack of plywood sheets each 4mm thick is 1.52m high. How many sheets are on the stack?	(A) 38 (B) 380 (C) 608 (D) 3800

36 The graph represents the distance a family were from Melbourne at any time on a car trip. On the way they stopped for a drink and later for lunch. Which point on the graph (A, B, C, D) represents the end of their lunch break?

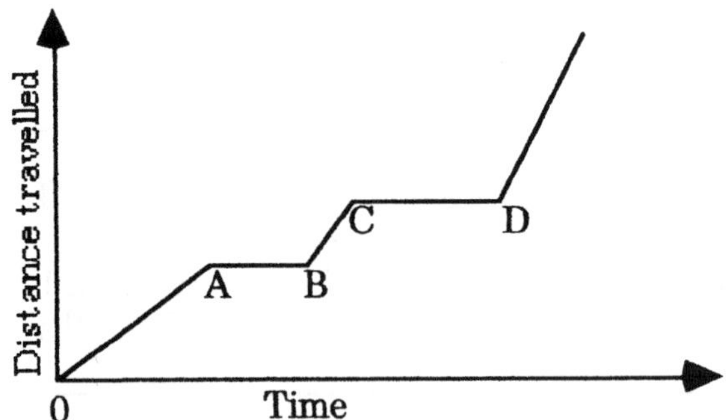

37 What transformation of the shaded triangle on Figure I will result in Figure II ?

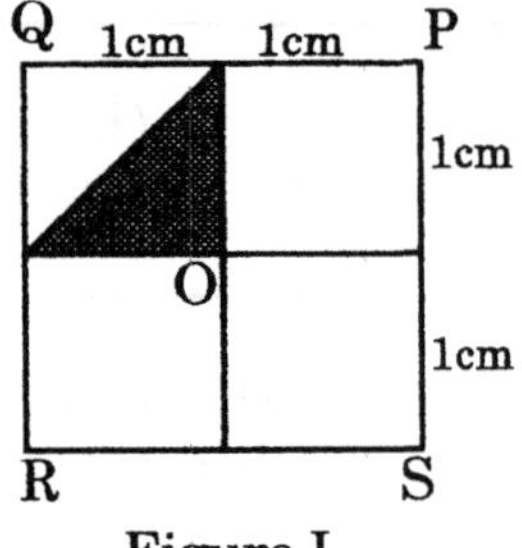

Figure I

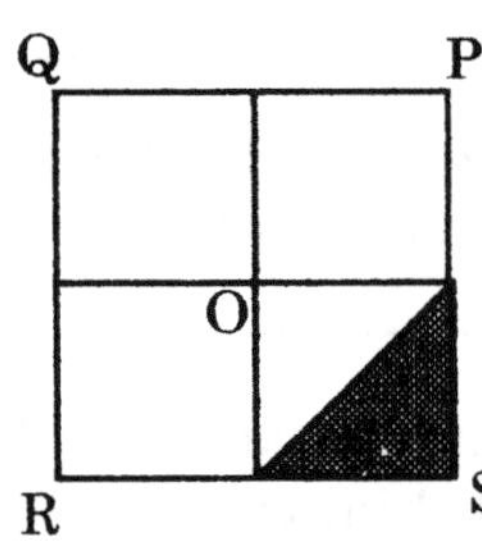

Figure II

(A) Reflecting it in the line PR
(B) Rotating it about O through an angle of 90° in a clockwise direction
(C) Rotating it about O through an angle of 180°
(D) Translating it 1cm to the right and then 1cm down

38 - 39 Four equal blocks are used to make each of the solids shown.

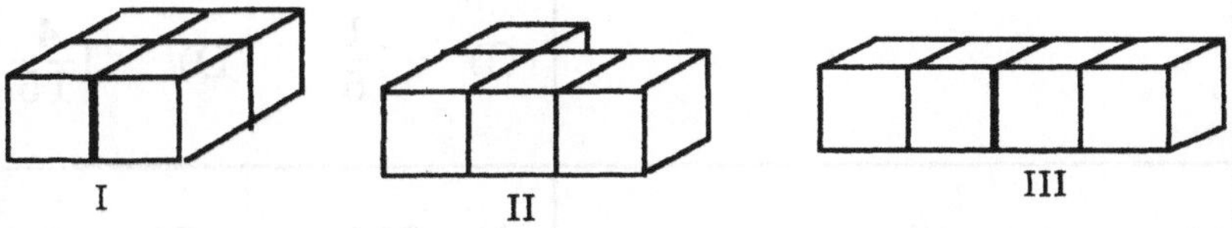

I II III

38 Which solid has the smallest surface area?

(A) I only
(B) II only
(C) III only
(D) they are all the same

39 Which solid has the largest volume?

(A) I only
(B) II only
(C) III only
(D) they are all the same

40 What is the most likely number needed to complete the following pattern?

1	2	3	4	5	6
0	4	18	48	100	

(A) 216
(B) 180
(C) 224
(D) None of these

PAPER 16

1	$\frac{3}{5} + 1\frac{1}{3} =$	(A) $1\frac{1}{2}$ (B) $1\frac{14}{15}$ (C) $1\frac{1}{5}$ (D) $1\frac{4}{15}$
2	Which of the following is incorrect?	(A) 215mm = 21cm 5mm (B) 899cm < 9m (C) 2km 42m > 2032m (D) 4m 65cm = 4065cm
3	43L is equal to .... mL	(A) 43 x 100 (B) 43 x 1000 (C) 43 x 10 (D) 43 x 10 000
4	How many eighths are there in $2\frac{1}{4}$?	(A) 24 (B) 18 (C) 12 (D) 9
5	A student finds the difference between 13^2 and 12^2. She takes the square root of her answer. Her final answer should be	(A) 25 (B) 2 (C) 1 (D) 5
6	Last year, John's height was 1.25m. This year his height increased by 24cm. What is his new height rounded off to the nearest metre?	(A) 1m (B) 2m (C) 1.5m (D) 2.5m
7	If the perimeter of the rectangle shown is 16.4m, what is its length? 240cm	(A) 14m (B) 11.5m (C) 5.8m (D) 4.8m
8	In the number pattern, what is the most likely missing number? 0.01, 0.29, 0.57, ... , 1.13	(A) 0.95 (B) 1.01 (C) 0.85 (D) 0.71
9	$6936 is divided equally among 4 workers. How much will each worker receive? Give your answer to the nearest ten dollars.	(A) $1730 (B) $1740 (C) $1700 (D) $1750

10	The teacher said to multiply by 6 and then add 5. I thought she said to divide by 6 and then subtract 5. I got 13 as my answer. The correct answer was	(A) 293 (B) 653 (C) 23 (D) 503
11	What percentage of the figure is shaded?	(A) 82.5% (B) 62.5% (C) 38.5% (D) 50%
12	A real estate agent was paid a commission of 3% for selling a unit for $120 000. How much was her commission?	(A) $36 (B) $360 (C) $3600 (D) $36 000
13	Half the difference of 18 and 6 is divided by twice their sum. The answer is	(A) $\frac{1}{8}$ (B) 8 (C) 2 (D) $\frac{1}{2}$
14	On a plan of a house to the scale of 1 : 100, a room measures 5.4cm. The actual length of the room is	(A) 5.4m (B) 54m (C) 0.54m (D) 4.6m
15	How many of the triangles can be made from the L - shaped figure?	(A) 4 (B) 6 (C) 8 (D) 10

16 - 17 The pie chart shows how Molly spent her pocket money on an excursion. 16 What fraction of her pocket money was spent on ice cream?	(A) $\frac{2}{3}$ (B) $\frac{8}{9}$ (C) $\frac{2}{9}$ (D) $\frac{1}{6}$
17 What was the sector angle for drinks?	(A) 60° (B) 90° (C) 120° (D) 150°
18 Pork costs $4.50 per kilogram and beef costs $5.40 per kilogram. How much altogether will Bob have to pay for 2.6kg of pork and 1.5kg of beef?	(A) $11.70 (B) $7.50 (C) $14.85 (D) $19.80
19 A particular brand of dress is sold at 3 for $69 during a sale. With $410 what is the greatest number of these dresses that you can buy?	(A) 16 (B) 17 (C) 18 (D) 19
20 A hiker walks 12km at 4km/h, rests for 1h, then walks the remaining distance of 6km at 3km/h. What was the average speed for the hike?	(A) 3km/h (B) 3.5km/h (C) 3.6km/h (D) 4km/h
21 If 32kg of butter is mixed with 8kg of sugar, what percentage of the mixture is butter?	(A) 20% (B) 25% (C) 50% (D) 80%
22 Gerald scored 84% in a test consisting of 25 questions. How many questions did he answer wrongly?	(A) 4 (B) 8 (C) 6 (D) 10
23 The diagram shows a balance. Each box marked X weighs the same. The other boxes have their weights in kilograms marked on them. The weight of a box marked X would be	(A) 7kg (B) 11kg (C) 3kg (D) $3\frac{2}{3}$kg

24	In a classroom the desks were arranged in a certain number of rows and the same number of columns. If you sat in Liam's desk there were two columns of desks to the left and three columns to the right. If 5 of the desks had no students sitting at them, the number of students in the class would be	(A) 20 (B) 11 (C) 30 (D) 31
25	Each of the symbols □ and Δ stand for numbers - each box represents the same number as does each triangle. | □ - Δ | □ + Δ | □ | Δ | $□^2 + Δ^2$ | | 7 | 19 | | | | By completing the table above, the value of $□^2 + Δ^2$ must be	(A) 205 (B) 410 (C) $102\frac{1}{2}$ (D) 676
26	When 7200 is resolved into prime factors, the number of 2's will be	(A) 3 (B) 4 (C) 5 (D) 6
27	A car travels from Town P to Town Q at 90km/h. If it takes 6 hours, how long would it take at 120km/h?	(A) $4\frac{1}{2}$ hours (B) 8 hours (C) 18 hours (D) 4 hours
28 - 29	P, Q, R,S are four points on a line such that Q, R, S are to the right of P. PQ = 3.4cm, PR = 2.6cm, PS = 4.9cm.	
28	Between which two nearest points does Q lie?	(A) P and R (B) P and S (C) R and S (D) none of these
29	If O is the midpoint of PQ, what is the length of OS?	(A) 4.0cm (B) 3.2cm (C) 2.3cm (D) 1.7cm
30	A ship originally travelling South-East changes direction to West. Through what angle does it turn?	(A) 45° (B) 90° (C) 135° (D) 150°
31	$4^2 - 10 \times 0.34 =$	(A) 2.04 (B) 12.6 (C) 4.6 (D) None of these

32	If 1 litre of paint covers 16 square metres and 150 square metres of wall are to be painted, then the number of 4 litre tins needed is	(A) 2 (B) 3 (C) 10 (D) $2\frac{1}{2}$
33	A city bus runs approximately every 17 minutes on a roughly circular route for tourists. The bus operates from 9:30a.m. to 5p.m. every day. How many complete trips does the bus do in a day?	(A) 25 (B) 26 (C) 27 (D) 28
34	Nina is to take 10mL of medicine every 3 hours from a bottle containing 300mL . If she starts taking the medicine on Tuesday at 11a.m, on which day does she finish the medicine?	(A) Sunday (B) Thursday (C) Friday (D) Saturday
35	If V is the number of vertices, F the number of faces, and E the number of edges for the solid shown, find the value of V + F - E.	(A) 2 (B) 12 (C) 18 (D) 4
36	A rectangular prism is 24cm x 4cm x 3cm. What is the smallest number of such prisms which can be used to build a cube? 24cm 4cm 3cm	(A) 8 (B) 6 (C) 24 (D) 48

37 - 38 The sketch shows two walls of a room 5m x 4m x 3m in which there is a door 2m x 1m and a window 1m x 0.8m. (There are no other doors or windows in the room.) 1m 3m 1m 2m 0.8m 5m 4m 37 The area of the 4 walls (excluding the door and window) is	(A) $51.2m^2$ (B) $56.8m^2$ (C) $52.2m^2$ (D) $52.8m^2$
38 The floor is to be tiled using tiles 25cm square. The number of tiles needed (not allowing for breakages) is	(A) 240 (B) 192 (C) 320 (D) 640
39 - 40 The symbol ϕ is made up to stand for "double and add 3". Hence, $\phi(4) = 2 \times (4) + 3 = 11$ and so on. 39 The value of $\phi\,(3\frac{1}{2})$ would be	(A) 7 (B) $9\frac{1}{2}$ (C) $10\frac{1}{2}$ (D) 10
40 If $\square$ stands for a number, and $\phi(\square) = 17$, then $\square$ must stand for	(A) 7 (B) $6\frac{1}{2}$ (C) $7\frac{1}{2}$ (D) 6

FURTHER MATHEMATICS TESTS

PAPER 17

1	75kg as a fraction of 1 tonne is	(A) $\frac{3}{40}$ (B) $\frac{3}{4}$ (C) $\frac{4}{3}$ (D) $\frac{40}{3}$
2	The number of axes of symmetry for this pattern is 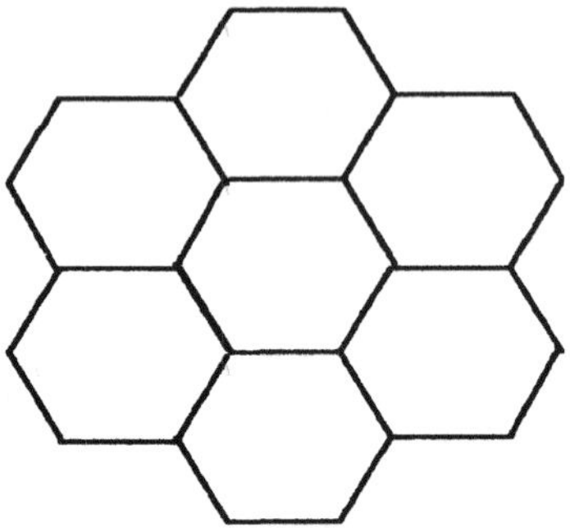	(A) 2 (B) 4 (C) 6 (D) 12
3	Fourteen children of a group of forty two drink apple juice everyday. What fraction of the group does not drink apple juice daily?	(A) $\frac{2}{3}$ (B) $\frac{1}{3}$ (C) $\frac{3}{5}$ (D) $\frac{3}{7}$
4	When a certain number is divided by 4, the remainder is 2 and the quotient is 17. The number must have been	(A) 70 (B) 25 (C) 34 (D) None of these
5	The price of a novel is $9.85. William wants to buy 4 of these novels with a $50 note. How much change will he receive?	(A) $39.40 (B) $89.40 (C) $10.60 (D) $11.60
6	What is the most likely number needed to complete the following pattern? 3, 7, 16, 32, 57, ☐	(A) 103 (B) 83 (C) 93 (D) None of these
7	There are 20 lamp posts equally spaced on a straight road. If the distance between the first and third lamp post is 21.4m, what is the distance between the 1st and the last lamp post?	(A) 214m (B) 203.3m (C) 428m (D) 406.6m

8	Find the perimeter of the triangle. 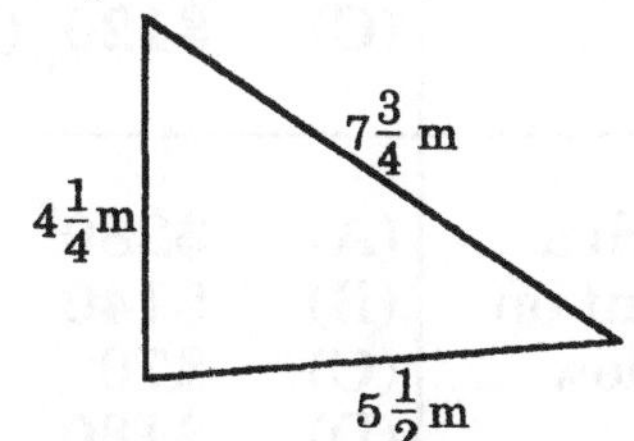	(A) $16\frac{3}{4}$ m (B) $15\frac{1}{2}$ m (C) $17\frac{1}{4}$ m (D) $17\frac{1}{2}$ m
9	A tank is $\frac{5}{12}$ full. If 30L of water is needed to make it half full, what is the capacity of the tank?	(A) 150L (B) 180L (C) 240L (D) 360L
10	Divide the difference between 42.8 and 10.56 by 2.6. The answer when corrected to the nearest whole number is	(A) 1 (B) 10 (C) 12 (D) 6
11	Peaches are sold at 5 for $1.75 while oranges cost 20% less. How much will 20 oranges cost?	(A) $7 (B) $6.50 (C) $6 (D) $5.60
12	Which one of the following gives the largest amount?	(A) 2% of $600 (B) 6% of $500 (C) 4% of $300 (D) 10% of $200
13	Two 4-digit numbers are formed from the digits 6, 1, 2, 7. The first has the digits in ascending order. The second has them in descending order. The difference of the two 4-digit numbers is	(A) 5445 (B) 5634 (C) 6354 (D) 4535
14 - 16	The sector graph shows how a person spends his weekly income. 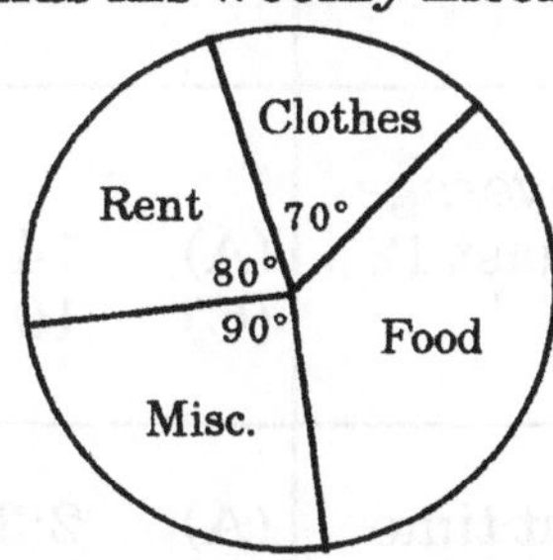	
14	What percentage of the total amount is spent on Food?	(A) $33\frac{1}{3}$% (B) 25% (C) $66\frac{2}{3}$% (D) 120%

15	If the total weekly expenditure is \$540, what amount is spent on Rent?	(A) \$80 (B) \$240 (C) \$120 (D) \$160
16	The weekly expenditure was changed in such a way that the difference in the amounts spent on Miscellaneous and Clothes was now \$35. How much was now spent on Rent?	(A) \$280 (B) \$140 (C) \$70 (D) \$180
17	A box contains a total of 200 buttons coloured green or black. There are 40 more green buttons than black ones. What percentage of the buttons in the box is black?	(A) 20% (B) 30% (C) 40% (D) 50%
18	Mr. Wood's yearly income is \$40 000. His yearly expenditure is between 80% to 90% of his income. He saves the rest of the income. Which of the following could not possibly be his weekly savings?	(A) \$68 (B) \$78 (C) \$108 (D) \$128
19	The smallest number which is divisible by all the numbers from 1 to 10 inclusive is	(A) $2^3 \times 3^3 \times 5 \times 7$ (B) $2^2 \times 3^3 \times 5 \times 7$ (C) $2^3 \times 5^2 \times 7$ (D) $2^3 \times 3^2 \times 5 \times 7$
20	What digit must be placed in the box so that the 4-digit number 3☐47 is divisible by 9	(A) 1 (B) 3 (C) 4 (D) 6
21	A pendulum swings back and forth 5 times in 2 seconds. How many times would it swing back and forth in 1 minute 50 seconds?	(A) 550 (B) 275 (C) 250 (D) 325
22	The average of 21 test scores is 18. The average of the first 8 is 17 and the average of the last 12 is 19. The 9th score is	(A) 14 (B) 17 (C) 19 (D) 21
23	The hands of a clock are at 12 noon. What time will it be when the minute hand has travelled through 540°?	(A) 2:30pm (B) 1:30pm (C) 5pm (D) 2:15pm

24 The diagram shows a balance. Each box marked X weighs the same. The other boxes have their weights in kilograms marked on them. X X X X 7	19 X X The weight of a box marked X would be	(A) 6kg (B) 13kg (C) 2kg (D) $4\frac{1}{3}$kg
25 The perimeter of a square field is 1200m. What is its area?	(A) 36ha (B) 6ha (C) 12ha (D) 9ha
26 A plane flew 200km in 25 minutes. What is its speed in km/h?	(A) 480 (B) 540 (C) $83\frac{1}{3}$ (D) 600
27 A photograph 8cm x 9cm is mounted on paper 9cm x 10cm. What fraction of the area of the paper is unused?	(A) $\frac{1}{4}$ (B) $\frac{3}{4}$ (C) $\frac{1}{5}$ (D) $\frac{4}{5}$
28 The sketch shows a box containing cans, each 20cm high and radius 6cm. If there are 2 rows of these cans, what are the dimensions of the box?	(A) 30cm x 18cm x 20cm (B) 30cm x 18cm x 40cm (C) 60cm x 36cm x 20cm (D) 60cm x 36cm x 40cm
29 A piece of tape 3m long is wrapped exactly once (without overlap) around the box shown. What is the length of the box? 55cm 15cm Diagram not drawn to scale	(A) 95cm (B) 270cm (C) 135cm (D) 80cm

30 Express a speed of 20m/s in km/h.	(A) 60km/h (B) 64km/h (C) 90km/h (D) 72km/h
31 - 32 A block of marble is 80cm x 48cm x 32cm. It is to be cut up into equal cubes. 32cm 80cm 48cm 31 What is the largest possible edge of a cube?	(A) 4cm (B) 8cm (C) 16cm (D) 32cm
32 How many cubes of side 4cm can be cut from this block?	(A) 96 (B) 160 (C) 240 (D) 1920
33 - 34 The sketch shows a solid made up of cubes of side 1cm. 33 How many cubes form the solid?	(A) 22 (B) 24 (C) 34 (D) 30
34 What is the surface area of the solid, (excluding the base)?	(A) $74cm^2$ (B) $60cm^2$ (C) $64cm^2$ (D) none of these
35 If 1cm of rain fell on a rectangular roof 30m x 20m and all of the water ran off into a tank, by how many cubic metres will the contents of the tank increase?	(A) 0.6 (B) 6 (C) 60 (D) 600

36	Four cities P, Q, R and S lie in a straight line, in that order. If P is 75km distant from S; R is 40km distant from P, and Q is 50km distant from S, then the distance from P to Q in kilometres is	(A) 25 (B) 30 (C) 35 (D) 40
37	The area of certain rectangle is Pcm^2. If each dimension is doubled, then the area of the new rectangle is	(A) 2 times Pcm^2 (B) 3 times Pcm^2 (C) 8 times Pcm^2 (D) 4 times Pcm^2
38	A rectangular prism has volume Vcm^3. If each dimension of the prism is trebled, then the volume of the new rectangular prism is	(A) 4 times V (B) 9 times V (C) 27 times V (D) 81 times V
39	A student finds the sum, difference, product and quotient of the numbers 21 and 7. She then adds all six numbers together. The result is	(A) 220 (B) 217 (C) 213 (D) 73
40	What is the most likely number needed to complete the following pattern?	(A) 32 (B) 28 (C) 35 (D) 37

2	5	1	3	6
5	26	2	10	

PAPER 18

1	Which set of fractions is in ascending order?	(A) $\frac{1}{4}, \frac{1}{5}, \frac{1}{6}, \frac{1}{7}$ (B) $\frac{1}{2}, \frac{3}{4}, \frac{5}{6}, \frac{9}{10}$ (C) $\frac{1}{10}, \frac{2}{9}, \frac{3}{7}, \frac{1}{3}$ (D) $\frac{1}{4}, \frac{1}{2}, \frac{2}{3}, \frac{3}{5}$
2	$1\frac{1}{3} \times 2\frac{5}{6} \times 2\frac{1}{4} =$	(A) $8\frac{1}{2}$ (B) $4\frac{5}{72}$ (C) $\frac{2}{17}$ (D) $6\frac{5}{12}$
3	Jason mixed 4.2L of water with 0.48L of syrup. He poured the mixture into 6 equal bottles. How much liquid was there in each bottle?	(A) 0.78L (B) 7.8L (C) 0.39L (D) 3.9L
4	Express 630m as a decimal of 1.5km.	(A) 0.24 (B) 0.63 (C) 0.42 (D) 0.36
5	Helen's mother is 50 years old. Helen's age is $\frac{2}{5}$ of her mother's. What will the sum of their ages be in 6 years' time?	(A) 50 years (B) 60 years (C) 76 years (D) 82 years
6	The height of a stack of bricks built with 25 rows of bricks is $2\frac{1}{2}$ m. If each brick has an equal thickness, what is the thickness of each brick?	(A) 1m (B) 10m (C) 10cm (D) 10mm
7	B, 18cm, A, C, D, 10cm In the figure above, ABC is an equilateral triangle. The perimeter of triangle BCD is 40cm. Express the perimeter of triangle ABC as a fraction of the perimeter of triangle ABD in its lowest terms.	(A) $\frac{9}{13}$ (B) $\frac{5}{26}$ (C) $\frac{7}{13}$ (D) $\frac{9}{26}$

8	A pile of 40 books of the same kind was 92.8cm high. A few books were removed from top of the pile and the height of the remaining books was 81.2cm. How many books were removed?	(A) 2 (B) 3 (C) 4 (D) 5
9	Which one of the following has a different value from the other three?	(A) 0.35 (B) $\frac{35}{50}$ (C) 35% (D) $\frac{7}{20}$
10	If you toss two dice, which of the following is more likely?	(A) You get exactly one "6" (B) You get at least one "6" (C) You get two "6"'s (D) All of the above are equally likely
11	Only one of the following equals 91. Which is it? (A) 10 x 10 - 10 + 10 ÷ 10 (B) 10 x 10 - (10 + 10) ÷ 10 (C) 10 x (10 - 10) + 10 ÷ 10 (D) (10 x 10 - 10 + 10) ÷ 10	
12	Which of the following is the largest?	(A) $\frac{8^8}{8 \times 8}$ (B) 8 x 8 x 8 x 8 (C) $8^8 \times 8^8$ (D) $(8 \div 8)^{88}$
13	The diagram shows a bucket and a marble. The bucket is filled with marbles of the same size and a student, knowing the size of the marble and bucket, has to give an estimate of the number of marbles in the bucket. Which of the following would be the best estimate?	(A) 50 (B) 1000 (C) 5000 (D) 10 000

14 David and Curtis were paid \$480 for painting a hall. David worked 9 hours and Curtis worked 15 hours. If they shared the money in proportion to the time each had worked, how much money did David get?	(A) \$300 (B) \$240 (C) \$200 (D) \$180
15 The cash price of a computer is \$1800. If I want to pay by instalments, I have to pay 15% more than the cash price. How much less do I to pay by cash than by instalments?	(A) \$540 (B) \$207 (C) \$504 (D) \$270
16 John picked 48 apples. After giving 12 apples to Lee and a few to Sam, he found that he had $\frac{2}{3}$ of the apples left. How many apples did Sam receive?	(A) 12 (B) 8 (C) 4 (D) 2
17 The entrance fees to a concert are as follows : Adults: \$35 per person Children: \$20 per person How much should be collected if 650 adults and 256 children are present on Friday whilst 725 adults and 484 children are present on Saturday?	(A) \$74 025 (B) \$62 925 (C) \$42 300 (D) \$48 125
18 Jenny has two \$50 notes, four \$20 notes, two \$10 notes, six \$5 notes and five \$2 coins. If she spends $\frac{3}{8}$ of her money, how much has she left?	(A) \$90 (B) \$150 (C) \$240 (D) \$80
19 What fraction of the whole figure is unshaded?	(A) $\frac{1}{4}$ (B) $\frac{3}{16}$ (C) $\frac{5}{12}$ (D) $\frac{13}{16}$
20 What number must be taken from five thousand and twenty three to give one thousand four hundred and seventy five?	(A) 3458 (B) 3548 (C) 3728 (D) 3755
21 10% of Ken's monthly income is the same as 5% of Terrence's. If Ken earns \$2500 per month, what is Terrence's monthly income?	(A) \$3000 (B) \$4000 (C) \$5000 (D) \$6000

22 There are four bells which ring at intervals of 3, 5, 6 and 7 minutes respectively. The bells ring simultaneously at 1p.m. The next time they ring together will be

(A) 4:30p.m.
(B) 1:30p.m.
(C) 7:00p.m.
(D) 3:30p.m.

23 There are four children in a family. Simon is 4 years older than Rachel and 5 years younger than Stephen. Judy is 2 years younger than Rachel. How old will Judy be when Stephen is 20 years?

(A) 31 years
(B) 11 years
(C) 9 years
(D) 29 years

24-25 Using the letters P, Q and R

24 how many 3 letter code words can be made if each letter is used once only in each code word?

(A) 6 (B) 3
(C) 9 (D) 27

25 how many 3 letter code words can be made if the letters can be used more than once in each code word?

(A) 6 (B) 12
(C) 24 (D) 27

26 What is the most likely number needed to complete the following pattern?

4	1	3	8	7	9
10	1	6	36	28	

(A) 64 (B) 54
(C) 45 (D) 48

27 The 4-digit number 1 □□ 1 is a perfect square. The numbers in the boxes, in order, could be

(A) 4 and 4
(B) 7 and 2
(C) 6 and 8
(D) 9 and 0

28 $0.63 \div 9 + 0.2^3 =$

(A) 0.708 (B) 1.3
(C) 0.15 (D) 0.078

29 - 31 An ant walks clockwise around the perimeter of a square of side 12cm.

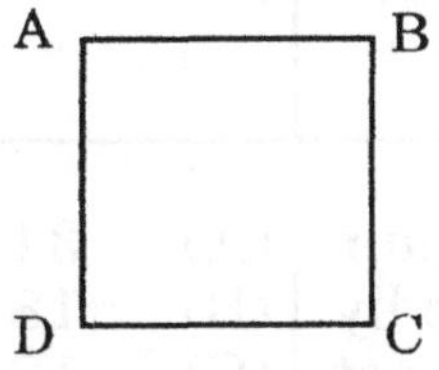

29 It walks from A to B at 1cm/s and from B to C at 2cm/s. The time taken to walk from A to C in seconds is

(A) 8 (B) 18
(C) 16 (D) 4

30	The ant walks from C to D at 3cm/s and from D to A at 4cm/s. The time taken to walk from C to A in seconds is	(A) 7 (B) $3\frac{3}{7}$ (C) $6\frac{6}{7}$ (D) 2
31	The average speed in cm/s of the ant walking completely around the square from A to A is	(A) $1\frac{23}{25}$ (B) $4\frac{4}{5}$ (C) $2\frac{1}{2}$ (D) 12
32	The average of 3 masses is 18kg and the average of another 5 masses is 14kg. What is the average of the 8 masses?	(A) 16kg (B) 15.5kg (C) 16.5kg (D) 124kg
33	If 8 men can plough a field in 9 hours, how long would it take 18 men at the same rate?	(A) 2 hours (B) $\frac{8}{9}$ hours (C) 4 hours (D) $\frac{9}{8}$ hours
34	What is the greatest number of coins weighing 22g each can be made from 1kg of alloy?	(A) 40 (B) 44 (C) 45 (D) 46
35	Each of the following numbers uses the digits 2, 3 and 4. Which is the smallest?	(A) 2×3^4 (B) 3×2^4 (C) $(3 + 2)^4$ (D) $(3 + 4)^2$
36	The product of the digits of a 5-digit number is 0. Which of the following must be true?	(A) All the digits are 0 (B) Most of the digits are 0 (C) At least one of the digits is 0 (D) The digits include a 2 and a 5
37	In a group of 44, an election was held for captain. There were two candidates. If Abed beat Colin by 18 votes, then the number of votes that Colin got was	(A) 31 (B) 18 (C) 13 (D) 26

38 There are 27 red marbles, 15 blue marbles and 18 yellow marbles in a box. What percentage of the marbles in the box are red?	(A) 65% (B) 55% (C) 45% (D) 35%
39 Two wooden rectangles are 7cm long and 11cm long respectively. They are placed lengthwise so that the length of their overlap is 5cm. A [diagram] B Diagram not drawn to scale From end to end, i.e. from A to B, their length in centimetres would be	(A) 13 (B) 18 (C) 23 (D) None of these
40 The diagram shows two rectangles. The first rectangle is 5cm wide and 9cm long. The other rectangle is 11cm long and 2cm wide. They are glued together to form the shaded figure shown in the second diagram. 5, 9, 2, 11 The area of the shaded figure in cm^2 is	(A) 5 x 9 + 2 x 11 + 10 (B) 5 x 9 - 2 x 11 - 10 (C) 5 x 9 - 2 x 11 + 10 (D) 5 x 9 + 2 x 11 - 10

PAPER 19

This PAPER contains some Challenge Questions

1	The perimeter of a rectangle is 94cm and the length is 3cm more than the breadth. The area of the rectangle in square centimetres is	(A) $\frac{94^2}{2}$ (B) 94 x 91 (C) 550 (D) 275

2 A rectangle is cut into two equal parts as shown :

Which of the following statements is true ?

(A) the sum of the perimeters of the two new rectangles equals the perimeter of the original rectangle.

(B) The area of the original rectangle is greater than the sum of the areas of the two new rectangles.

(C) The perimeter of the original rectangle is greater than the sum of the perimeters of the two new rectangles.

(D) The sum of the perimeters of the two new rectangles is greater than the perimeter of the original.

3 The travel graph shows a trip made by Sally.
Which of the following is not true?

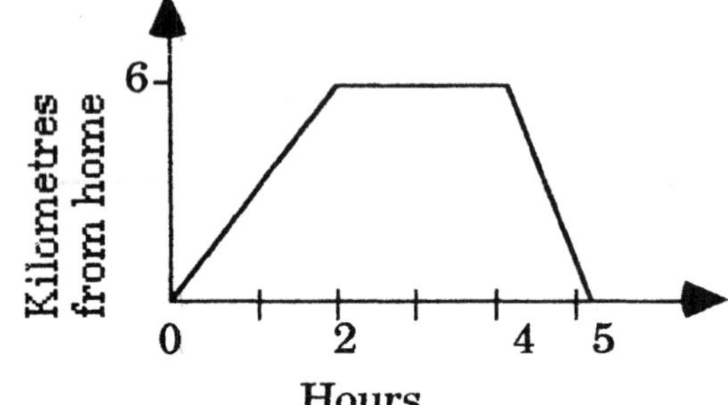

(A) Sally went 12km altogether.

(B) Sally started the journey from home and ended the journey at home.

(C) Sally must have travelled up a flat topped hill and down the other side.

(D) Sally stopped for 2 hours during the trip.

4 Two identical pieces of cardboard have a slit in them as shown.

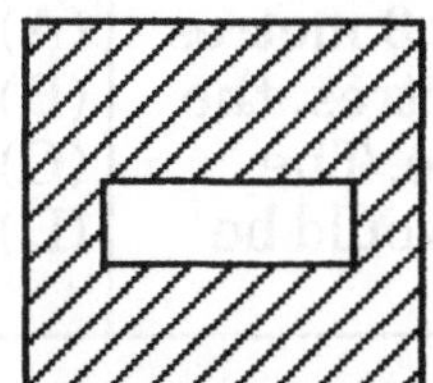
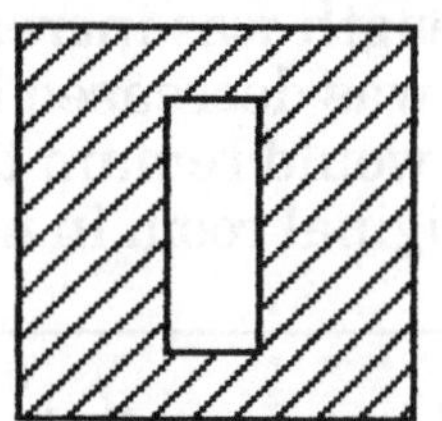

One piece of cardboard is placed on top of the other with the slits still in the directions shown in the sketch. A light is shone through the combined slit onto a wall. The shape on the wall will be

(A) 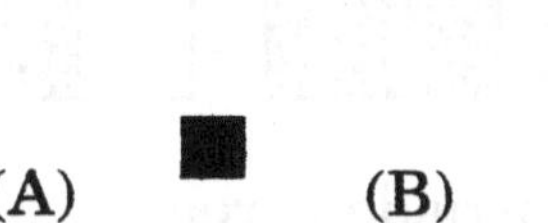 (B)  (C)  (D) 

5 - 6 All of the solids shown have the same height.

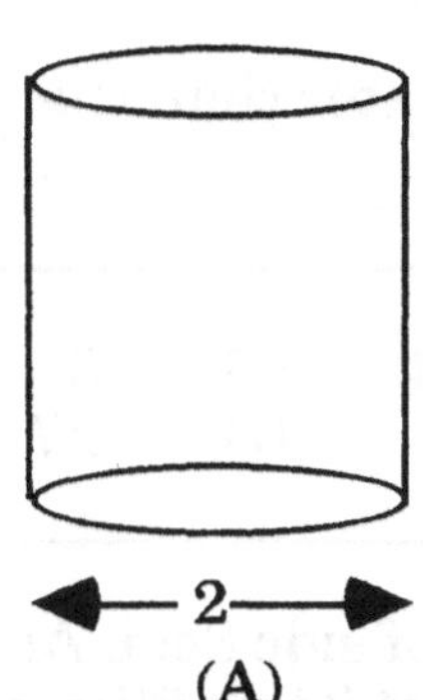

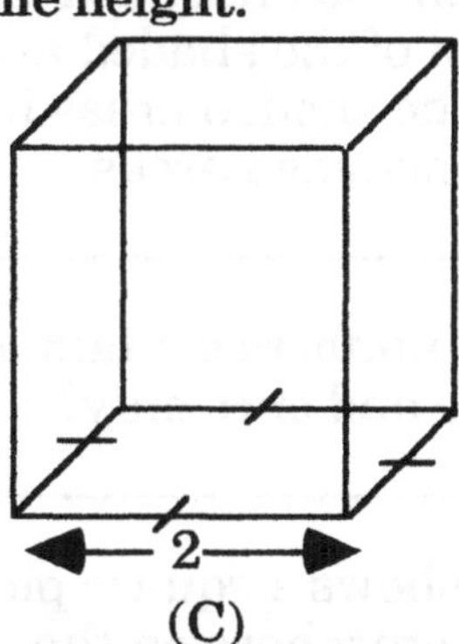

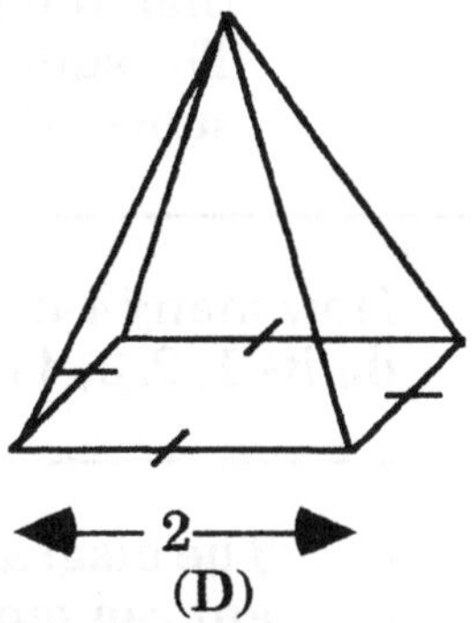

5	Which of them would have the greatest volume?	(A) (B) (C) (D)
6	Which would have the smallest volume?	(A) (B) (C) (D)
7	A student counts all the digits used to number the pages of a book. If 615 digits were used, starting at page 1, the number of pages in the book will be	(A) 615 (B) 241 (C) 328 (D) None of these

8 - 9 A button is placed on a number line at the 100 mark. A die is tossed 6 times and the results added together. If the answer is an odd number, then the button is moved that number of places to the right. If the answer is even, then the button is moved that number of places to the left.

8	The greatest number that the counter can end up on is	(A) 135 (B) 146 (C) 105 (D) 101
9	The smallest number that the counter can end up on is	(A) 94 (B) 64 (C) 100 (D) 36

10	The length of a room is 3 metres more than its breadth. If the length was increased by 3 metres and the breadth was decreased by 2 metres, the area of the room would remain the same. The breadth of the original room in metres would be	(A) 4 (B) 8 (C) 10 (D) None of these

11 The following squares are all the same size. One of the following statements is false.

 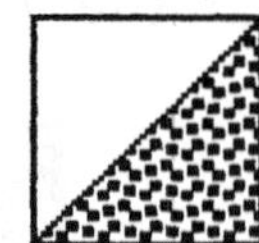 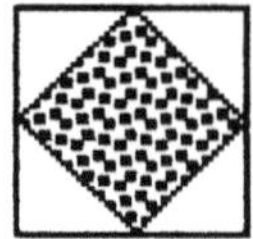 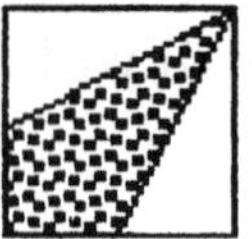 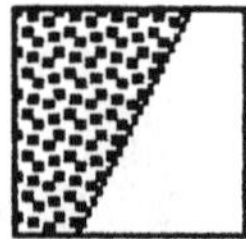 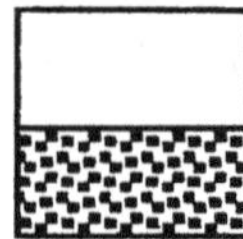 

(A) In each diagram the shaded area equals the unshaded area.
(B) The shaded area in each diagram is not less than the unshaded area.
(C) The sum of all the unshaded areas in all the diagrams is not greater than the sum of the shaded areas.
(D) The sum of the shaded areas in all the diagrams is greater than the sum of the unshaded areas.

12	How many 4-digit numbers contain all of the digits 1, 2, 3, 4 once and once only?	(A) 24 (B) 3 (C) 9 (D) 27

13 - 15 The diagram shows a square piece of cardboard PQRS of side 8cm. An ant can move anywhere on the cardboard. An example of its position is the point T where its distances from the edges of the square are shown by dotted lines

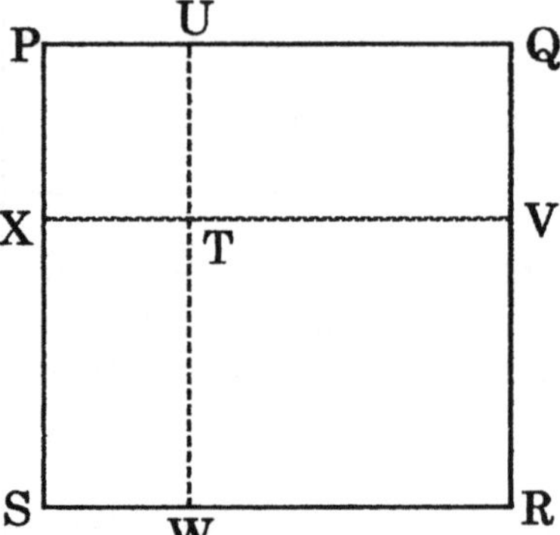

13 Which of the following statements is true?
(A) No matter where the ant moves, the sum of its distances from the sides PS and QR is 16cm.
(B) No matter where the ant moves, the sum of its distances from the four sides of the square is 32cm.
(C) No matter where the ant moves, the sum of its distances from sides PQ and SR equals the sum of its distances from PS and QR.
(D) No matter where the ant moves, the sum of its distances from PQ and PS is always less than the sum of its distances from SR and QR.

14 As the ant moves, the square is divided into 4 rectangles as shown, whose areas change.

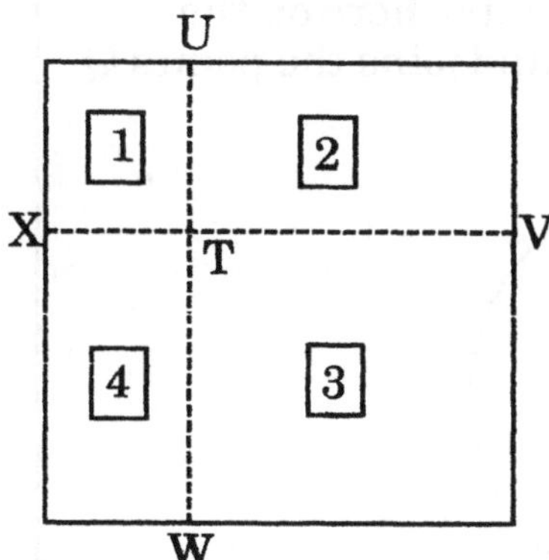

Which of the following statements is true?

(A) No matter where the ant moves, the sum of the areas [1] and [2] is $32cm^2$.

(B) No matter where the ant moves, the sum of areas [1] [2] [3] and [4] is $64cm^2$.

(C) No matter where the ant moves, the sum of the areas [2] and [3] is greater than the sum of the areas [1] and [4].

(D) No matter where the ant moves, the sum of the areas [1] and [3] is not less than the sum of the areas [2] and [4].

15 As the ant moves, the perimeter of each of the rectangles [1] [2] [3] and [4] changes. Which of the following statements is true?

(A) As the ant moves, the perimeter of [3] is greater than [1]

(B) As the ant moves, the sum of the perimeters of [2] and [3] is greater than the sum of the perimeters of [1] and [4].

(C) As the ant moves, the sum of the perimeters of [1] and [3] equals the sum of the perimeters of [2] and [4] .

(D) As the ant moves, the sum of the perimeters of [1] [2] [3] and [4] equals the perimeter of PQRS.

16 The diagram shows a circle of diameter 12cm. The fixed point Q is 14cm from the centre O of the circle. A girl puts a point P anywhere on the circle. A girl puts a point P anywhere on the circumference of the circle and joins the points Q and P with a straight line. P, O, Q The shortest possible length of QP in centimetres is	(A) 8 (B) 20 (C) 14 (D) 6
17 Three ice cubes, of sides 1cm, 2cm and 3cm respectively, are placed in a cylinder whose base is $4.5cm^2$. After the ice cubes have melted, the height of the water in centimetres in the cylinder will be	(A) 12 (B) 9 (C) 8 (D) 16
18 The teacher said to divide by 8 and then add 11. I thought he said to multiply by 8 and subtract 11. I got 437. The correct answer was	(A) 75 (B) 53 (C) 18 (D) 77

19 I have 17 marbles which I wish to place into 12 cups. Which of the following statements must be true?

(A) There must be at least one marble in each of the cups.
(B) One cup must contain at least 5 marbles.
(C) Five cups must contain more than one marble.
(D) There must be at least 2 marbles in at least 1 cup.

20 A man died in 1793 aged 75 years and his son died in 1789 aged 41 years. How old was the father when the son was born?	(A) 30 (B) 34 (C) 35 (D) None of these

21 What is the most likely number needed to complete the following pattern?

(A) 13
(B) 19
(C) 16
(D) 21

22 One hundred and fourteen buttons are divided into three piles. Together, the first and second piles contain 76 buttons. Together, the second and third piles contain 85 buttons. The second pile contains

(A) 47
(B) 29
(C) 38
(D) 41

23 Approximately, how many working days would be needed to count one million dollars if you counted 60 dollars each minute and worked a 24 hour day?

(A) 12 (B) 5
(C) 144 (D) 20

24 In the following addition, each letter stands for a number and different letters do not stand for the same number.

```
  C R O S S +
    R O A D S
  D A N G E R
```

Which of the following statements is true?

(A) R stands for an odd number
(B) D A N G E R is a prime number
(C) C and R are both less than 5
(D) D A N G E R must be less than 200 000

25

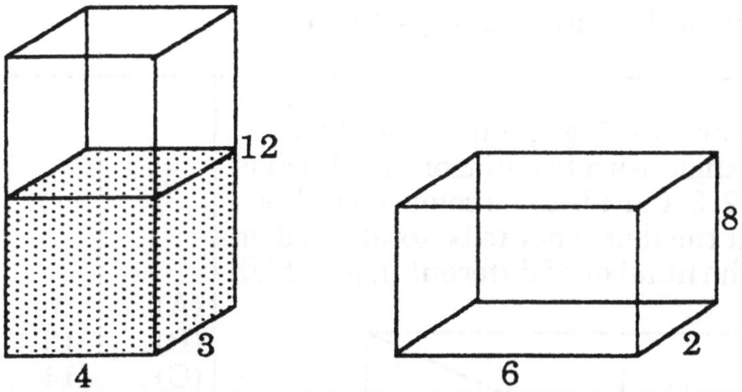

In the sketch above, the vase on the left has dimensions 4cm x 3cm x 12cm. It is half-full of water. The water is emptied into the vase at the right, which measures 6cm x 2cm x 8cm. The depth of the water in the second vase will be

(A) 2
(B) 4
(C) 6
(D) 8

26 - 27

P Q

A car is travelling a straight road towards two suburbs, P and Q, which are 6km apart. As he travels, the driver adds his distance from P to his distance from Q. Initially, the car is 42km to the left of P.

26 If the driver continues to calculate the sum of the distances until he is 42km past Q, the greatest sum in kilometres he will obtain will be

(A) 84 (B) 90
(C) 96 (D) 132

27 The least value of the sum of his distances in kilometres will be

(A) 6 (B) 52
(C) 48
(D) None of these

28 The diagram shows a wire frame formed by 16 squares, each of side 2cm. A spider at P wants to go to Q. It can walk along the wires only in the directions of East and North, i.e. to the right and up. Which of the following statements is false?

Q

P

(A) The distance the spider travels is the same no matter what path is followed
(B) The distance travelled depends on the path taken.
(C) The distance that the spider travels in the northerly direction, equals the distance that it travels in the easterly direction
(D) The length of the path is always 16cm

29 The diagram shows the flag of a new country. It is to be a rectangle with two diagonals. Each of the triangles 1, 2, 3 and 4 is to be coloured either red or green, but the flag is not to be totally red or totally green. The number of different flags which can be made is

1
2 4
3

(A) 8
(B) 12
(C) 14
(D) 16

30 When something takes a long time to learn we say that it has a "long learning curve". Which of the following graphs would best represent a long learning curve?

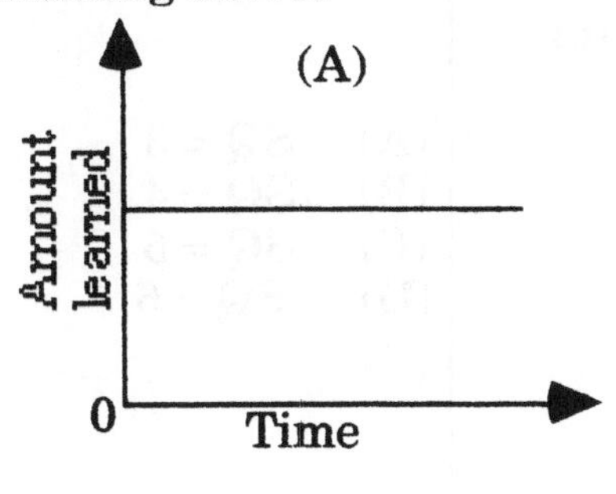

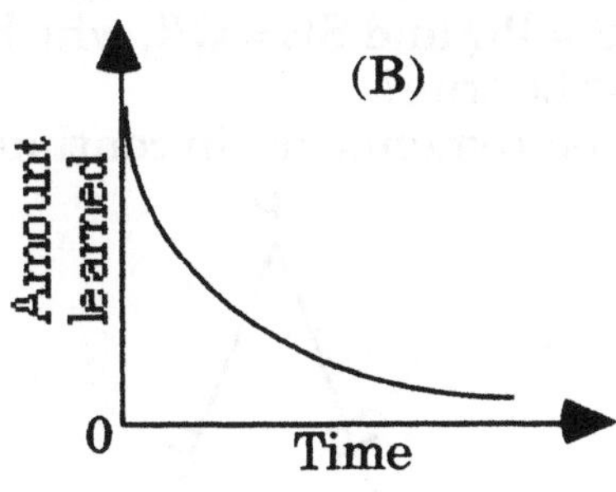

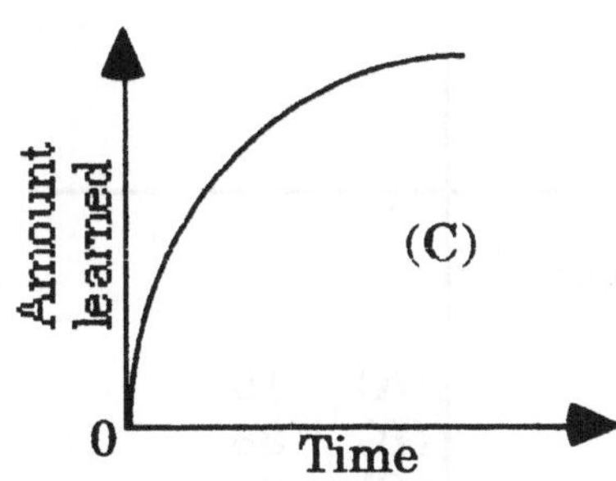

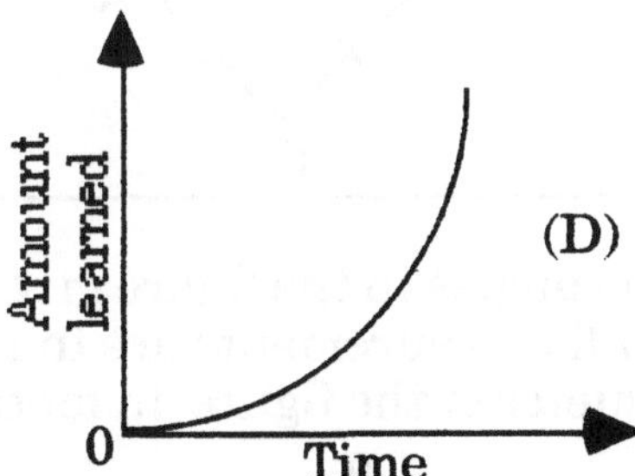

31 A vase contains 3 red roses, 3 yellow roses and 5 white roses. One rose is picked at random. Which of the following statements is false?

(A) You are less likely to pick a red rose than you are to pick a white rose.
(B) You have a smaller chance of picking a white rose than not picking a white rose.
(C) Picking a red rose is not as likely as picking a yellow rose.
(D) You are certain to get a red, yellow or white rose.

32 The end of a solid is like an H as shown in the first sketch. When the solid is seen from either side, it looks like the second sketch. The number of blocks needed to build the solid will be:

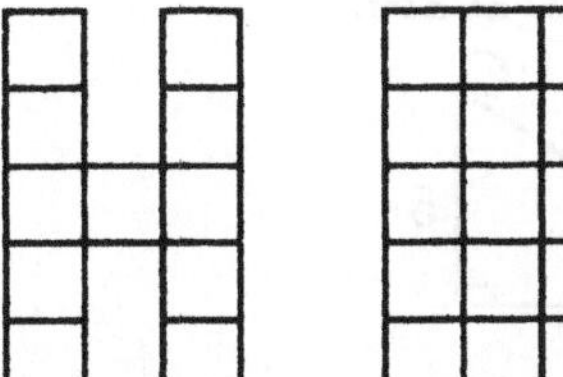

(A) 11
(B) 36
(C) 61
(D) None of these

33 The perimeter of the kite PQRS is 26cm.
The perimeter of the triangle PQS is 21cm.
The perimeter of the triangle RQS is 15cm.
If PS = PQ and SR = QR, which of the following must be true?
(Measurements are in centimetres)

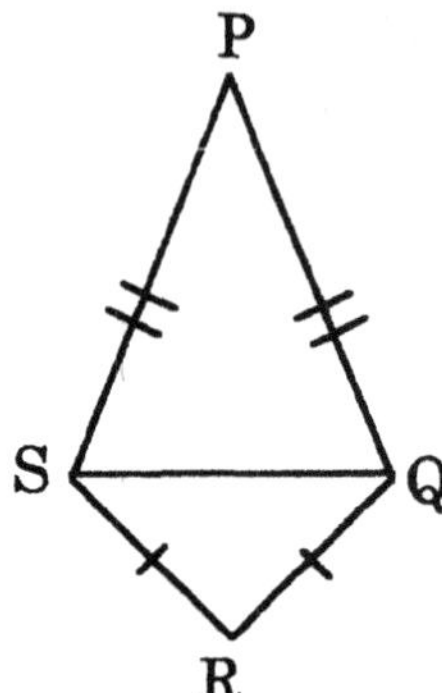

(A) SQ = 3
(B) SQ = 4
(C) SQ = 5
(D) SQ = 6

34 All the angles in the figure given are right angles and all measurements are in metres. The perimeter of the figure in metres is

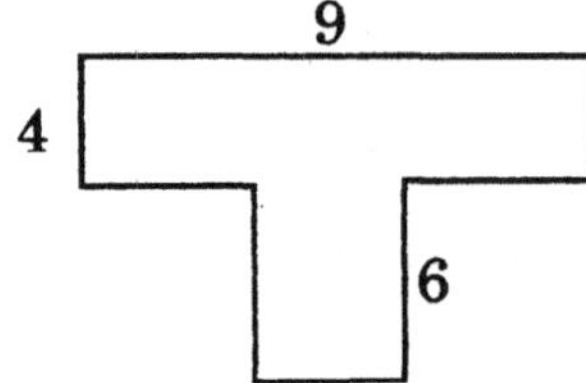

(A) 19
(B) 38
(C) 24
(D) there is not enough information to tell

35 In which of the following figures are the diagonals perpendicular to one another?
(A) Kites, rhombuses and rectangles
(B) Parallelograms and rhombuses
(C) Squares, rhombuses and kites
(D) Rhombuses and rectangles

36 A steel ball is placed at the top of a slope. The ball rolls down the slope. In which of the following cases would the ball be travelling fastest, just before it leaves the slope? [Diagrams are not drawn to scale.]

(A)
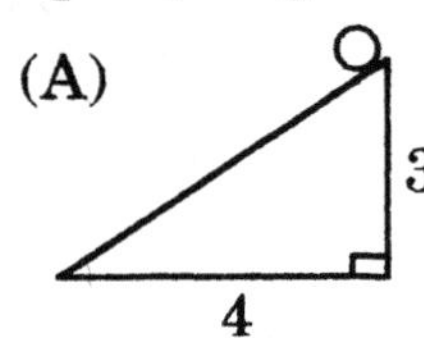

(B)
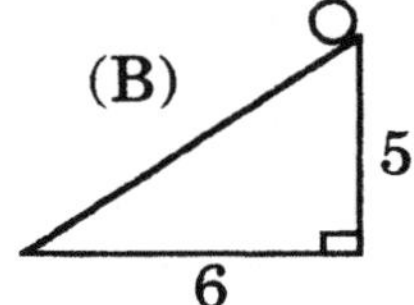

(C)
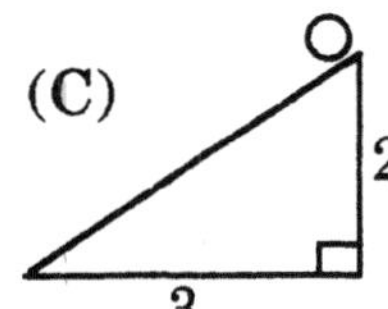

(D)
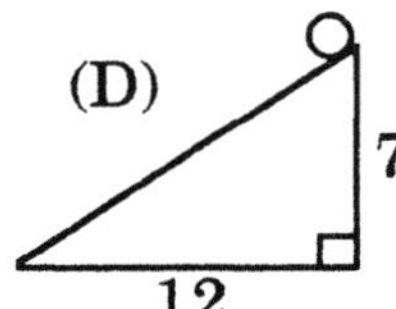

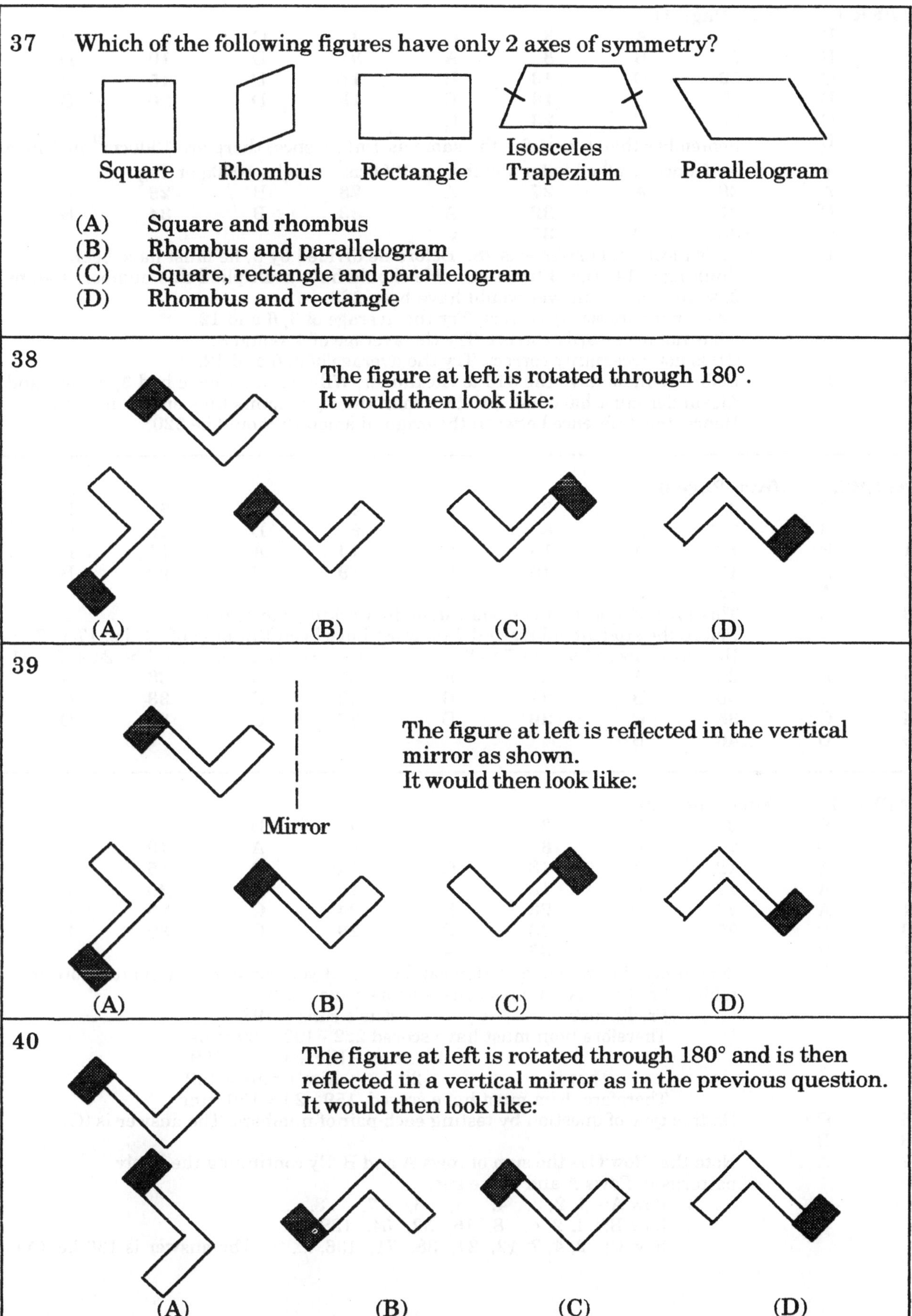

37 Which of the following figures have only 2 axes of symmetry?

(A) Square and rhombus
(B) Rhombus and parallelogram
(C) Square, rectangle and parallelogram
(D) Rhombus and rectangle

38 The figure at left is rotated through 180°. It would then look like:

(A) (B) (C) (D)

39 The figure at left is reflected in the vertical mirror as shown. It would then look like:

(A) (B) (C) (D)

40 The figure at left is rotated through 180° and is then reflected in a vertical mirror as in the previous question. It would then look like:

(A) (B) (C) (D)

PAPER 1 **(from Page 1)**

1	B	**2**	A	**3**	C	**4**	C	**5**	B
6	B	**7**	B	**8**	A	**9**	D	**10**	D
11	C	**12**	D	**13**	B	**14**	D	**15**	B
16	D	**17**	C	**18**	D	**19**	D	**20**	B
21	C	**22**	C	**23**	C				

24 D Remember that 1cm^3 of is the same as 1mL. Hence, there are 1000cm^3 in 1 litre. And since 1cm^3 of water weighs 1g, 1000cm^3 weighs 1000g or 1kg.

25	A	**26**	A	**27**	A	**28**	B	**29**	B
30	C	**31**	C	**32**	A	**33**	B	**34**	B
35	C	**36**	D	**37**	C				

38 D Lee's incorrect answer was 48. Before he divided by 3, he must have been thinking of 144 (i.e. 3 times 48). He should have multiplied this number, 144, by 2, so the correct answer would have been 288.

39 B (A) is not necessarily correct. Try the average of 3, 6 and 12.
(C) is not necessarily correct. Try the average of 4, 6 and 8.
(D) is not necessarily correct. Try the average of 3, 6 and 12.

40 B Before William gave the $10 to Alexander, William must have had $10 more and Alexander must have had $10 less than the equal sums they ended up with. Hence, the difference between the original amounts would be $20.

PAPER 2 **(from Page 6)**

1	C	**2**	C	**3**	D	**4**	D	**5**	B
6	B	**7**	B	**8**	C	**9**	D	**10**	C
11	B	**12**	A	**13**	D	**14**	A	**15**	B
16	A	**17**	C	**18**	B	**19**	B	**20**	B
21	A	**22**	C						

23 A This type of question is easiest done by working backwards:
Before the student subtracted 7 he must have been thinking of 24, i.e. 17 + 7.
He was supposed to add 7 to 24, so that the correctanswer would be 24 + 7 = 31

24	B	**25**	A	**26**	A	**27**	B	**28**	A
29	C	**30**	B	**31**	B	**32**	C	**33**	C
34	C	**35**	C	**36**	D	**37**	C	**38**	D
39	D	**40**	B						

PAPER 3 **(from Page 12)**

1	B	**2**	C	**3**	B	**4**	D	**5**	C
6	D	**7**	B	**8**	B	**9**	A	**10**	D
11	A	**12**	A	**13**	C	**14**	B	**15**	C
16	A	**17**	D	**18**	A	**19**	A	**20**	D
21	A	**22**	D	**23**	B	**24**	C	**25**	B
26	D	**27**	D	**28**	C	**29**	C	**30**	D
31	B	**32**	D	**33**	A				

34 D Sometimes this type of question can be done if you write the information out in a table: Fred's runs + Ivan's runs + Kim's runs = 222
Fred's runs + Kim's runs = 193
Therefore Ivan must have scored 222 - 193 = 29 runs
Ivan's runs + Kim's runs = 159
That is 29 + Kim's runs = 159
Therefore Kim must have scored 159 - 29 = 130 runs

35 C Do this type of question by testing each pair of numbers. The answer is (C)

36 B

37 A Note that Row C is the sum of rows A and B. By continuing the likely patterns in Rows A and B, we get:
Row A: 1, 2, 3, 4, 5, 6, 7, 8, ...
Row B: 1, 2, 4, 8, 16, 32, 64, 128, ...
Row C: 2, 4, 7, 12, 21, 38, 71, 136, ... The answer is 136 i.e. (A)

38	A	Imagine that we had several of these buckets. In 6 minutes the first tap would fill 2 buckets and the second tap would fill 3 buckets. That is : 5 buckets would be filled in 6 minutes by the two taps. Therefore 1 bucket would be filled in 6 ÷ 5 minutes = 1 minute 12 seconds. OR In 1 minute, the first tap would fill $\frac{1}{3}$ of the bucket and the second tap would fill $\frac{1}{2}$ of the bucket. Together, the 2 taps would fill $\frac{1}{3} + \frac{1}{2} = \frac{5}{6}$ of the bucket. Hence, in $\frac{6}{5}$ minutes, the 2 taps together would fill $\frac{6}{5} \times \frac{5}{6} = 1$ bucket - i.e. the bucket would be filled in $\frac{6}{5}$ minutes, i.e. 1 minute 12 seconds.
39	B	This is a "think of a number" type question. Dividing the original number by 3 and multiplying the answer by 6 is the same as doubling the original number. When we divide this answer by 2 we would get back to the original number. Adding 5 to this result and then subtracting the original number would leave us with an answer of 5, no matter what the original number was. The answer is (B)
40	D	This is done by "guess and check". We might try 9 x 10 x 11 = 990 which is too large. Then we might try 7 x 8 x 9 = 504 which is too small. But we soon find that the correct answer would be 8 x 9 x 10 = 720. The sum of these numbers would be 8 + 9 + 10 = 27 so the correct answer must be (D) None of these.

PAPER 4 (from Page 17)

1 D **2** C **3** D **4** C **5** D
6 B **7** D **8** D **9** C **10** D
11 B
12 A

In the first 2 minutes the car travels 1000m
That is, it travels 500m in 1 minute
That is, it travels 500 x 60 m in 1 hour
That is, it travels $\frac{500 \times 60}{1000}$ km in 1 hour
That is, it travels at 30km/h

13 D **14** C **15** B **16** C **17** D
18 B

Remember that you do not get the correct answer by adding the times together!
Mr. Chan would have to pay the car parking bill on the first day, i.e. \$30 and on the second day he would have to pay \$15. The answer is \$30 + \$15 = \$45 i.e. (B)

19 B **20** B **21** D **22** C **23** B
24 A **25** B
26 A

6 persons pay a total of \$174 ∴ each person pays \$174 ÷ 6 = \$29
Thus, there are 377 ÷ 29 = 13 persons in the group

27 D **28** C **29** C **30** A **31** B
32 C **33** C
34 C

The mass of an object remains the same no matter where the object is.
The weight varies from place to place - it may even be zero as you see in space walks on television.
Here we were asked for the mass of Kim and Fido which must be 48kg on Earth or on the Moon. The information about their weights is not needed.

35 D

Testing each answer in mL:

First glass	Then the second glass must hold	Total
Try 200	200 - 80 = 120	320
Try 140	140 - 80 = 60	200
Try 100	100 - 80 = 20	120
Try 180	180 - 80 = 100	280

36 B **37** A **38** B **39** A **40** A

PAPER 5 (from Page 22)

1 A **2** C **3** D **4** A **5** D
6 C **7** B **8** B **9** D **10** B
11 C **12** A **13** B **14** D **15** A
16 C **17** C **18** B **19** B **20** C

21	B	22	B	23	C	24	C	25	B
26	D	27	C	28	B	29	A	30	C
31	D	32	B	33	C	34	B		

35 D

By subtraction, we find the other sides as shown in the diagram.
Thinking of the block as two rectangular prisms, we find that the surface area is $2(5 \times 4 + 3 \times 4 + 2 \times 4 + 1 \times 3) = 86 \text{ cm}^2$

36 A Note that each of the given answers involves the number 53. Six persons could receive \$53 each. The seventh would then receive \$53 plus an extra \$6. Therefore there are seven lots of \$53 plus an extra \$6. Note that if the answers had been
(A) $(7 x 218 + 6) (B) $(6 x 218 + 7)
(C) $(13 x 218) (D) $(13 x 218 + 7) the answer would still be (A). In this case the first 6 persons would receive \$218 each.

37	D	38	A	39	C	40	B

PAPER 6 (from Page 27)

1	D	2	A	3	A	4	D	5	C
6	A	7	B	8	C	9	D	10	B
11	B	12	A	13	A	14	D	15	C
16	A	17	C	18	C	19	D	20	D
21	B	22	D	23	B	24	C	25	A
26	C	27	B	28	B	29	C	30	B
31	C								

32 B Drawing a vertical line at the point X divides the rectangle PQRS into equal rectangles. The shaded area is then seen to be $\frac{3}{4}$ of the total area.

33 C We can get the answer by trial and error: If the base BC were 12cm, then the other sides would add to 30cm and since they are of equal length, each would be 15cm. But BC is to be longer than the other sides. Hence BC cannot equal 12cm. Similarly BC cannot be 14cm.
But if BC = 16cm, then AB and AC are each 13cm. Hence, the answer is (C)

34	B	35	C	36	B	37	D	38	A

39 C In order to leave a remainder of 1 when divided by 6, a number must be of the form $6 \times \square + 1$, where the $\square$ stands for any whole number - try it! And to leave a remainder of 1 when divided by 7, the number must be able to be written as $7 \times \square + 1$.
So, to leave a remainder of 1 when divided by both 6 and 7, the number must be able to be written as $6 \times 7 \times \square + 1$. The only answer given which can be written like this is $6 \times 49 \times 5 + 1 = 6 \times 7 \times \boxed{35} + 1$. Therefore (C) is the answer.

40 A

PAPER 7 (from Page 33)

1	C	2	C	3	B	4	C	5	B
6	C	7	A	8	B	9	C	10	D
11	A	12	D	13	C	14	A		

15 A Note that the information in the first sentence is not used at all.

16	D	17	D	18	B	19	B	20	D

21 B Since 7 is 2 units away from 9, we want numbers which are 6 units away from 9 - namely 3 and 15. Don't forget to draw a diagram to help you see the answer.

22	B	23	C	24	C	25	A	26	A

27	B	**28**	A	**29**	C	**30**	B	**31**	B
32	B	**33**	A	**34**	B	**35**	D	**36**	C
37	C	**38**	D						

39 C We can find the answer by making a table and testing the possible solutions. Ignoring dollar signs we have:

Say that Jan receives	Then Shmuel receives	And Tom receives	Total
18	18	3 x 18 = 54	18 + 18 + 54 ≠ 72
54	54	3 x 54 = 162	54 + 54 + 162 ≠ 72
14.40	14.40	3 x 14.40 = 43.20	14.40 + 14.40 + 43.20 = 72

40 D

PAPER 8 (from Page 38)

1	D	**2**	C	**3**	C	**4**	D	**5**	A
6	A	**7**	B	**8**	D	**9**	D	**10**	A
11	C	**12**	D	**13**	D	**14**	A	**15**	B
16	D	**17**	C	**18**	A	**19**	C	**20**	D
21	D	**22**	A						

23 D We could find the smallest number by dividing each number by 2,3,4,5,6,7 and 8 in turn. We can save time by noticing that if the number is divisible by both 2 and 3, then it must be divisible by 6. Hence we don't have to divide by 6 but by 2,3,4,5, 7 and 8. Also if the number is divisible by 8, we don't have to divide by 2 and 4. Hence we need only divide by 3, 5, 7 and 8. Thus the answer is (D).

24 B The food will last 12 men for 28 days
Therefore the food should last one man for 12 x 28 days
Thus the food should last 14 men for $\frac{12 \times 28}{14}$ = 24 days. Hence, the answer is (B).

25 D

26 D Since the stream is flowing at 4km/h and the girl can row at 3km/h, it is as though the stream is still and the girl is rowing at 7km/h. Using the basic formula: speed = $\frac{\text{distance}}{\text{time}}$ we find that the journey will take 2 hours.

27	D	**28**	B	**29**	C	**30**	B	**31**	A
32	B	**33**	C	**34**	A	**35**	A	**36**	B
37	D								

38 D We need to find the largest number which divides exactly into 224 and 576, i.e. we need the highest common factor of 224 and 576. By testing each of the four numbers in turn, we find that the answer is (D).

39 D Try making a table and using trial and error - $120 was chosen as our first trial.

Cost of Phone x 3	Cost of radio x 2	Total	Comment
$120 x 3 = $360	$80 x 2 = $160	$520	Too small
$600 x 3 = $1800	$400 x 2 = $800	$2600	Too large
$300 x 3 = $900	$200 x 2 = $400	$1300	Correct

Or, we could test the solutions in turn: (Amounts are in dollars)

Cost of Radio	Cost of Phone	Total
100	100 x $\frac{3}{2}$ = 150	100 x 2 + 150 x 3 = 650
400	400 x $\frac{3}{2}$ = 600	400 x 2 + 600 x 3 = 2600
300	300 x $\frac{3}{2}$ = 450	300 x 2 + 450 x 3 = 1950
200	200 x $\frac{3}{2}$ = 300	200 x 2 + 300 x 3 = 1300

40 C The units for height are metres.

Day	Climbs	Maximum height reached before slipping back	Slips back	Total height at end of day
1	3	3	1	2
2	3	2 + 3 = 5	1	4
3	3	4 + 3 = 7	1	6
4	3	6 + 3 = 9	1	8
5	3	8 + 3 = 11	1	10
6	3	10 + 3 = 13	1	12
7	3	12 + 3 = 15	1	14
8	3	14 + 3 = 17	1	16
9	3	16 + 3 = 19	1	18
10	3	18 + 3 = 21		

Hence, the greatest height reached is 21m. The answer is (C)

PAPER 9 **(from Page 43)**

1 B **2** C **3** B **4** B **5** B
6 B **7** B **8** D **9** B **10** B
11 A **12** D **13** B **14** A **15** D
16 B
17 D After a 10% discount, you would pay 90% of the original price for each skirt.
i.e. 90% corresponds to \$54
∴ 10% corresponds \$54 ÷ 9 = \$6
∴ 100% corresponds to \$60, i.e. each skirt originally cost \$60
∴ the original price for 10 skirts is \$600
18 A **19** C
20 D If 3 plings = 8 plungs,
then 12 plings = 32 plungs.
Therefore, 5 plangs = 32 plungs, since 5 plangs = 12 plings
Therefore, 25 plangs = 5 x 32 = 160 plungs. The answer is (D)
21 A
22 A Remember that if a whole number is to be divisible by 7, then it must have a factor of 7. Hence, if □ and Δ represent whole numbers, a number of the form 7 x □ x Δ would always be divisible by 7. And, a number of the form 7 x □ x Δ + 2 would leave a remainder of 2 when divided by 7. Similarly, a number of the form 11 x □ x Δ + 2 leaves a remainder of 2 when divided by 11. And a number of the form 13 x □ x Δ + 2 leaves a remainder of 2 when divided by 13. Thus a number of the form 7 x 11 x 13 + 2 leaves a remainder of 2 when divided by 7, 11 and 13. Hence, the answer is (A)
23 D **24** B
25 C Note that 920B.C. is before 776B.C. The answer is (C)
26 B **27** A **28** A **29** B **30** B
31 A **32** C **33** B **34** C **35** D
36 C There are 219 lots of 582 (plus 171) in 127 629. Hence we could subtract 582 from 127 629 a total of 219 times. The answer is (C)
37 C If 75 men take 12 days, then 75 x 12 men should take 1 day.
Hence, $\frac{75 \times 12}{20}$ men should take 20 days. The answer is (C)
38 D Try each of the numbers in turn.
39 B **40** C

PAPER 10 **(from Page 48)**

1 C **2** B **3** A **4** B **5** B
6 B **7** D **8** D **9** A

10	C	Do this type of question by testing the given solutions - don't forget the meaning of the word "exceeds" as it occurs quite often. If you find $\frac{1}{3}$ of 60 and then find $\frac{1}{4}$ of 60, you'll find that $\frac{1}{3}$ of 60 exceeds $\frac{1}{4}$ of 60 by 5. The answer is (C)							
11	D	12	D	13	A	14	C	15	C
16	D	17	A						
18	C	The dots will be on the ground together every 12 metres. Since there are 8 twelves in 100 the answer seems to be 8. But don't forget the first one. The answer is (C)							
19	B								
20	B	This question has too much information. We don't need to know the sum of the numbers. The answer is (B)							
21	C	22	C	23	C	24	B	25	C
26	B	27	A	28	C	29	D	30	B
31	A	32	C	33	A	34	B	35	D
36	C	37	A	38	D	39	C	40	C

PAPER 11 (from Page 54)

1	D	2	C	3	C	4	D	5	B
6	D	7	A	8	C	9	D	10	C
11	B	12	B	13	C	14	A	15	B
16	B	17	A	18	A	19	B	20	C
21	D	22	C	23	D	24	B		
25	B	Adding 5, 7, 9, ... to the respective terms 7, 12, 19, ... , gives the next term.							
26	C	27	A	28	C	29	C	30	B
31	B	32	B	33	D	34	B	35	C
36	B	37	D	38	C	39	C	40	A

PAPER 12 (from Page 59)

1	A	2	D	3	A	4	B	5	D
6	C	7	D	8	C	9	D	10	A
11	A	12	D	13	B	14	C	15	A
16	C	17	C	18	A	19	D	20	C
21	B	22	A	23	C	24	B	25	A
26	A	27	A	28	D	29	D	30	B
31	D	One square metre measures 100cm by 100cm. Hence it contains 100 x 100 = 10 000 square centimetres. And hence, in $24m^2$ there will be 24 x 10 000 = 240 $000cm^2$							
32	C	33	C						
34	A	5kg per 1 litre is the same as 5000g per 1000mL i.e. 5g per 1mL i.e. 5g per $1cm^3$ [Remember that $1cm^3$ = 1mL]							
35	D	36	A	37	C	38	B	39	D
40	D	Method 1:We could do this question by testing the solutions given. Method 2: We could argue that $\frac{3}{4}$ of Grace's money has been spent, and so she is left with only $\frac{1}{4}$ of what she started with. Since this equals $12.50, she must have started with 4 x 12.50 = $50. The answer is (D)							

PAPER 13 (from Page 64)

1	C	2	A	3	C	4	C	5	A
6	A	7	C	8	D	9	C	10	C
11	D	12	D	13	B	14	D	15	C
16	A	17	C	18	B	19	D		
20	B	Each pair of numbers adds to 20.							
21	A	22	C	23	B	24	B	25	C
26	C	27	A	28	C	29	C	30	C
31	C	32	A	33	A	34	B	35	C
36	B	37	C						

38 B Ask yourself the question: what would be the weight of, say, 10 packets? Obviously the answer is 6kg. So we could fill more than 10 packets.
Try 15 packets. Their weight would be 9kg which is too much and therefore we must fill less than 15 packets.
Try 13 packets. Now we get 7.8kg and we would have only 0.2kg left over so we could not make an extra packet. Hence the answer is (B)

39 B Don't forget the formula: $\text{Speed} = \dfrac{\text{total distance travelled}}{\text{total time taken}}$
To use this formula we need to know how far the journey was. If you travel at 60km/h for $1\frac{1}{4}$ hours, you will have travelled 75km. And if you travel at 90km/h for $\frac{1}{4}$ of an hour, you will have travelled $22\frac{1}{2}$ km. So the total distance will be $97\frac{1}{2}$ km and you will have taken $1\frac{1}{2}$ hours. So your speed = $\dfrac{97\frac{1}{2}}{1\frac{1}{2}}$ = 65km/h.
[There is an easy way to simplify this fraction: just double the top and double the bottom giving $\frac{195}{3}$. This is like cancelling in reverse and it always works.]

40 D We could fill in a table to help, using the given ages for the older sister.
[Note: Ages are in years.]

Older sister's age	Younger sister's age	Difference in ages	$\frac{1}{7}$ of the sum of their ages
48	56 - 48 = 8	40	8
36	56 - 36 = 20	16	8
40	56 - 40 = 16	24	8
32	56 - 32 = 24	8	8

PAPER 14 **(from Page 70)**

1	D	**2**	C	**3**	B	**4**	C	**5**	B
6	D	**7**	D	**8**	A	**9**	B	**10**	B
11	D	**12**	A	**13**	A				

14 D Note that the mention of the two year period is not needed.

15 D Seven years ago, the sister would have been 15 - 7 = 8 years old. The man would then have been 2 x 8 = 16 years old.
Today the man must be 16 + 7 = 23 years old.

16 A It depends where we hold the ruler.

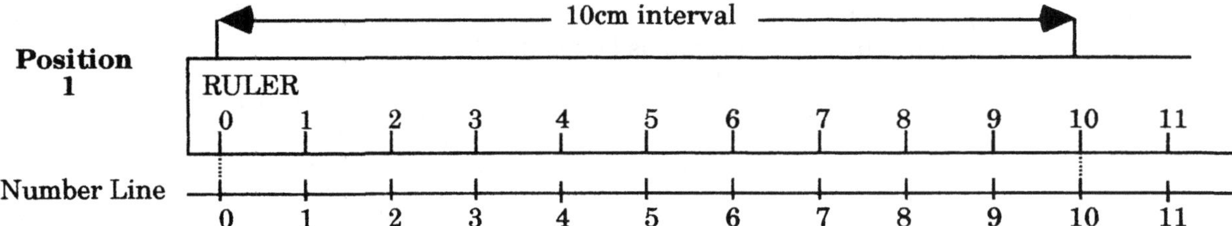

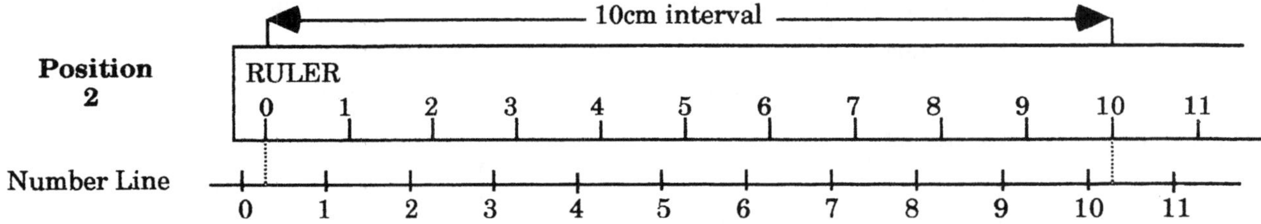

In Position 1, there are 9 numbers in the 10cm interval.
In Position 2, there are 10 numbers in the 10cm interval. The answer is 9.

17	D	**18**	D	**19**	B	**20**	C	**21**	D

22 C The second row is found by subtracting twice the number in the first row from 24 i.e. if you double the numbers in the first row and add them to the corresponding numbers in the second row, you always get 24. It often helps you to see the pattern if you write the numbers in the first row from smallest to largest first:

1	3	5	7	8	10
	18	14	10	8	4

Or you might see the pattern more easily if you include all the numbers from 1 to 10 :

1	2	3	4	5	6	7	8	9	10
		18		14		10	8		4

23 D **24** B **25** C **26** A **27** C
28 B **29** C **30** B **31** C **32** B
33 D **34** B **35** A **36** C **37** C
38 C **39** B **40** B

PAPER 15 **(from Page 76)**

1 D **2** A **3** A **4** C **5** B
6 C **7** A **8** C **9** C **10** D
11 D **12** C **13** D **14** C **15** D
16 B **17** C

18 D The first 9 pages would need 1 digit each: total = 9 digits. The pages from 10 to 99 - i.e. 90 pages would need 2 digits each: total = 180 digits
The pages from 100 to 256 - i.e. 157 pages would need 3 digits each:
total = 471 digits Altogether 9 + 180 + 471 = 660 digits. The answer is (D)

19 B If 15% of a sum of money is \$96 then 1% would equal $\$\frac{96}{15}$

Hence, 25% would equal $25 \times \$\frac{96}{15} = \$160$. [Note that we don't need to find the sum of money if we set the problem out in this way. By not cancelling fractions until the last step, we can often save a lot of time.]

20 C **21** B **22** B

23 C If the Large Number is divisible by 3 and 4, it will certainly be divisible by the lowest common multiple of 3 and 4, namely 12. The answer is (C)
[If the Large Number had been divisible by 8 and 10, the next smallest number that it certainly would be divisible by would be 40, i.e. the LCM of 8 and 10.]

24 D When there are 2 persons in the room there will be 1 handshake. If a third person enters the room, he must shake hands with the two persons in the room. Hence, there will be 1 + 2 = 3 handshakes altogether.
If a fourth person enters the room, she will have to shake hands with the three persons in the room. Hence, there will be 1 + 2 + 3 = 6 handshakes altogether.

25 D **26** D **27** C **28** D **29** B
30 C **31** A **32** D **33** C **34** B
35 B **36** D **37** D **38** A **39** D

40 B $1^2 \times 0 = 0$; $2^2 \times 1 = 4$; $3^2 \times 2 = 18$; $4^2 \times 3 = 48$; $5^2 \times 4 = 100$; $6^2 \times 5 = 180$

PAPER 16 **(from Page 82)**

1 B **2** D **3** B **4** B **5** D
6 A **7** C **8** C **9** A **10** B
11 B **12** C **13** A **14** A **15** B
16 C **17** C **18** D **19** B **20** A
21 D **22** A **23** C **24** D **25** A
26 C **27** A **28** C **29** B **30** C
31 B **32** B **33** B **34** D **35** A
36 D **37** A **38** C **39** D **40** A

PAPER 17 **(from Page 88)**

1 A **2** C **3** A **4** A **5** C

6 C The difference between the first 2 numbers is 4, between the next two numbers is 9, between the next two numbers is 16 and so on. That is the difference between the numbers is a perfect square. The answer is (C)

7	B	**8**	D	**9**	D	**10**	C	**11**	D
12	B	**13**	C	**14**	A	**15**	C	**16**	B
17	C	**18**	A						

19 D

Note that $1 = 1$ $2 = 2$ $3 = 3$ $4 = 2^2$ $5 = 5$ $6 = 2 \times 3$ $7 = 7$ $8 = 2^3$ $9 = 3^2$ $10 = 2 \times 5$

The smallest number must have a factor of 2^3, 3^2, 5 and 7.
Thus the smallest number is $2^3 \times 3^2 \times 5 \times 7$

20	C	**21**	B	**22**	A	**23**	B	**24**	A
25	D	**26**	A	**27**	C	**28**	D	**29**	C
30	D	**31**	C	**32**	D	**33**	D	**34**	C
35	B	**36**	A	**37**	D	**38**	C	**39**	A

40 D

The second number is one more than the square of the first.

PAPER 18 **(from Page 94)**

1	B	**2**	A	**3**	A	**4**	C	**5**	D
6	C	**7**	A	**8**	D	**9**	B	**10**	B
11	A	**12**	C						

13 B

If we placed one layer of marbles across the bottom of the bucket, there would be about 60 marbles. There appear to be about 15 layers of marbles to fill the bucket. So we would need about $60 \times 15 = 900$ marbles to fill the bucket.

14	D	**15**	D	**16**	C	**17**	B	**18**	B
19	D	**20**	B	**21**	C				

22 A

We must add the LCM of 3, 5, 6 and 7 - namely 210 minutes - to 1p.m, giving 4:30 p.m. The answer is (A)

23	C	**24**	A	**25**	D

26 C

$\frac{4 \times 5}{2} = 10$ $\frac{1 \times 2}{2} = 1$ $\frac{3 \times 4}{2} = 6$

$\frac{8 \times 9}{2} = 36$ $\frac{7 \times 8}{2} = 28$ $\frac{9 \times 10}{2} = 45$

27 C

We must test to see if the number 1441 is a perfect square. It is a good idea to study the following table of perfect squares:

$1^2 = 1$ $2^2 = 4$ $3^2 = 9$ $4^2 = 16$ $5^2 = 25$ $6^2 = 36$ $7^2 = 49$ $8^2 = 64$ $9^2 = 91$ $10^2 = 100$	We know that $30^2 = 900$ and $40^2 = 1600$, so if 1441 is a perfect square, its square root must lie between 30 and 40. Therefore, since 1441 ends in a 1, we know that if it has a square root, then this root must be 31 or 39 - since 31^2 and 39^2 both end in 1 - try them! Neither of them give 1441. We now try 1721. Using the same thinking, 41 and 49 are the only possibilities. Neither of them work. So we try 1681 and find that it is 41^2. Hence, the answer is (C)

28	D	**29**	B	**30**	A	**31**	A	**32**	B
33	C	**34**	C	**35**	B				

36 C

Remember that the question says "must be true". Statement (C) is the only one that must be true. Statements (A) and (B) might be true and statement (D) would not mean that the product was necessarily zero.

37 C Try each solution in turn:

Colin's vote	Therefore Abed's vote was	Total number of votes
31	31 + 18 = 49	80
18	18 + 18 = 36	54
13	13 + 18 = 31	44

38 C **39** A **40** D

PAPER 19 **(from Page 100)**

1 C **2** D **3** C **4** A **5** C
6 B
7 B

Number of pages using 1 digit = 9, namely pages 1 to 9 inclusive
Total So Far = 9 digits
Number of pages using 2 digits = 90, namely pages 10 to 99 inclusive.
Total So Far = 9 + 90 x 2 = 189 digits
Number of pages using 3 digits = 900, namely pages 100 to 999 inclusive
Total So Far = 189 + 900 x 3 = 2889 digits - which is too large.
We want only some of these 900 pages if the total is to be 615 digits.
We want the Total So Far = 189 + ❏ x 3 = 615
The number in the box must be 142. So we have

Number of pages using 1 digit = 9
Number of pages using 2 digits = 90
Number of pages using 3 digits = 142
Total number of pages = 9 + 90 + 142
= 241

8 A **9** B
10 D We can do this problem by testing the given answers:

Original breadth (m)	Original length (m)	Original area (m^2)	New breadth (m)	New length (m)	New area (m^2)
4	7	4 x 7 = 28	4 - 2 = 2	7 + 3 = 10	2 x 10 = 20
8	11	8 x 11 = 88	8 - 2 = 6	11 + 3 = 14	6 x 14 = 84
10	13	10 x 13 = 130	10 - 2 = 8	13 + 3 = 16	8 x 16 = 128

11 D **12** A **13** C **14** B **15** C
16 A The shortest length will occur when QP passes through the centre and P is between Q and O
17 C The volume of the first ice cube is $1 \times 1 \times 1 = 1\ cm^3$
The volume of the second ice cube is $2 \times 2 \times 2 = 8 cm^3$
The volume of the third ice cube is $3 \times 3 \times 3 = 27 cm^3$
The total volume of the ice cubes is $1 + 8 + 27 = 36 cm^3$
The formula for the volume of a cylinder is
Volume = area of the base times the perpendicular height*
We know that the volume of the water from the melted ice cubes is $36 cm^3$
So we have: 36 = 4.5 times perpendicular height
Hence, the perpendicular height = 36 ÷ 4.5 = 8. The answer is (C)
18 C Before I subtracted 11, I must have been thinking of 437 + 11 = 448.
Before I multiplied by 8, I must have been thinking of 448 ÷ 8 = 56.
If I had done what the teacher said I would have first divided 56 by 8 giving 7, and then added 11 giving 18. The correct answer is (C)
19 D
20 A The father was born in 1793 - 75 = 1718
The son was born in 1789 - 41 = 1748
Therefore the father was 1748 - 1718 = 30 when the son was born.
21 A 2 x 2 - (2 + 2) = 0
3 x 2 - (3 + 2) = 1
5 x 3 - (5 + 3) = 7
7 x 4 - (7 + 4) = 17
8 x 3 - (8 + 3) = 13
22 A First pile + second pile = 76
Second pile + third pile = 85
Adding: First pile + Twice the second pile + third pile = 161
But we know that First pile + second pile + third pile = 114
Therefore the second pile must contain 161 - 114 = 47

23 A In one day you would count \$60 in 1 minute, i.e. \$60 x 60 in 1 hour i.e. \$60 x 60 x 24 in 1 24-hour day

Hence, you would need $\frac{1\,000\,000}{60 \times 60 \times 24}$ days

i.e. $\frac{10\,000}{36 \times 24}$

i.e. approximately $\frac{10\,000}{40 \times 20}$

i.e. $\frac{100}{8} = 12.5$ days approximately

24 D Since R = S + S, R must be even since it is twice a number. Since R is even, D A N G E R is even and hence is not prime - since it is divisible by 2. If both C and R were less than 5, then they could not add to give a 2-digit number - even if you were carrying 1 from the previous column. D must be 1 since two 1-digit numbers cannot add to more than 19 (you may be carrying 1). Hence 100 000 < D A N G E R < 200 000

25 C The volume of water in the left vase will be $\frac{1}{2} \times 4 \times 3 \times 12 = 72\text{cm}^3$. The volume of water in the right vase will be 6 x 2 x H, where H is the depth of water. Obviously, 72 = 6 x 2 x H. Hence, H = 6cm

26 B The greatest sum will occur when the car is at the starting point or the finishing point. The greatest sum will therefore be 42 + 48 = 90km

27 A The least sum will occur when the car is between P and Q or at P or Q. Hence, the least sum is 6km.

28 B

29 C Each triangle can be either red or green i.e. there are 2 colour choices for each triangle. Therefore there are 2 x 2 x 2 x 2 = 16 different ways of colouring the flag. But this includes the possibilities of all triangles being red or all triangles being green - which is not allowed. Therefore there are 2 x 2 x 2 x 2 - 2 = 14 ways.

30 C Diagram A would mean that as time increases, the amount learned remains constant. Diagram B implies that as time increases, the amount learned decreases. Diagram C and D both imply that as time increases, the amount learned increases. But Diagram C means that as time increases, the amount learned in any time interval becomes smaller - as we gradually learn how to do the something. On the other hand, Diagram D implies that the amount we learn in any time interval gets larger and larger and, according to the diagram, shows no sign of levelling off. Hence, D could hardly represent the learning of something.

31 C We have 3 chances of picking a red rose and 5 chances of picking a white. Hence (A) is true. We have 5 chances of picking a white rose and 6 chances of not picking one. Hence (B) is true. There are 3 chances of picking a red rose and 3 chances of picking a yellow rose, i.e. the chances are equal. Hence (C) is false. Finally, you must get a red, yellow or white rose. So (D) is true.

32 D We are not sure - since we are not told - whether the one block in the centre of the H extends the full length of the solid. We can only be certain that the number of blocks used will be at least 25 + 25 + 1 = 51 and - if there are 5 centre blocks - there will be at most 25 + 25 + 5 = 55 blocks.

33 C 34 B 35 C

36 B We could redraw each diagram using different scales so that the base of each triangle was 12.

(A) 9, 12 (B) 10, 12 (C) 8, 12 (D) 7, 12

Hence, (B) can be seen to have the greatest slope and hence, the ball would be travelling fastest when it left this slope.

37 D 38 D 39 C 40 B

NOTES

NOTES

Coroneos Publications – Titles by Topic
BASIC SKILLS SERIES

ISBN	Item	Title	Author
Basic Skills Maths Tests			
9781862941076	42	Further Maths Tests for Selective Schools Scholarship Exams Yrs 5–8	Coroneos,An,Smith
9781862940697	49	Maths Tests for Selective Schools Scholarship Exams Yrs 5–8	Coroneos,An,Smith
Basic Skills Tests			
9781862941243	112	Further Selective Schools Scholarship Tests Multiple Choice Yrs 5–8	Peter Howard
9781862941359	125	Year/Grade 4 Tests (O.C.) multiple choice Yrs 4–5	Peter Howard
9781862940598	64	Selective Schools Scholarship Tests Yrs 5–8	Peter Howard
9781862940758	73	Basic Skills Test Yrs 3–8 - Suitable preparation for NAPLAN Tests	Peter Howard
9781862941403	144	Year 3 Basic Skills Test - Suitable preparation for NAPLAN Tests	Peter Howard
9781862941632	153	Year 5 Basic Skills Test - Suitable preparation for NAPLAN Tests	Peter Howard
NAPLAN* Format Practice Tests			
9781921565434	235	Language Conventions Year 3 - NAPLAN* Format Practice Tests	Don Robens
9781921565441	236	Language Conventions Year 5 - NAPLAN* Format Practice Tests	Don Robens
9781921565533	245	Language Conventions Year 7 - NAPLAN* Format Practice Tests	Don Robens
9781921565458	237	Writing Year 3 - NAPLAN* Format Practice Tests	Alfred Fletcher
9781921565465	238	Writing Year 5 - NAPLAN* Format Practice Tests	Alfred Fletcher
9781921565519	243	Writing Year 7 - NAPLAN* Format Practice Tests	Alfred Fletcher
9781921565526	244	Writing Year 9 - NAPLAN* Format Practice Tests	Alfred Fletcher
9781921565472	239	Reading Year 3 - NAPLAN* Format Practice Tests	Alfred Fletcher
9781921565489	240	Reading Year 5 - NAPLAN* Format Practice Tests	Alfred Fletcher
9781921565557	247	Reading Year 7- NAPLAN* Format Practice Tests	Alfred Fletcher
9781921565496	241	Numeracy Year 3 - NAPLAN* Format Practice Tests	Don Robens
9781921565502	242	Numeracy Year 5 - NAPLAN* Format Practice Tests	Don Robens
9781921565540	246	Numeracy Year 7 - NAPLAN* Format Practice Tests	Don Robens
Basic Skills General Ability			
9781862942868	176	Giant Book of General Ability Tests Yrs 5–8	Graeme Wilson
9781862940857	81	General Aptitude Tests for Selective Yrs 5–8	Peter Howard
9781862941724	154	IQ Examples Level 1 Yrs 3–4	Peter Howard
9781862941731	155	IQ Examples Level 2 Yrs 5–8	Peter Howard
9781862940727	70	Learn to Think Level 1 Yrs 2–7	Peter Howard
9781862940734	71	Learn to Think Level 2 Yrs 5–8	Peter Howard
Basic Skills Language and Maths			
9781862940635	50	Kindergarten	Peter Howard
9781862940475	51	Year 1 Language/Mathematics	Peter Howard
9781862940482	52	Year 2 Language/Mathematics	Peter Howard
9781862940499	53	Year 3 Language/Mathematics	Peter Howard
9781862940505	54	Year 4 Language/Mathematics	Peter Howard
9781862940512	55	Year 5 Language/Mathematics	Peter Howard
9781862940529	56	Year 6 Language/Mathematics	Peter Howard
9781862940765	74	Year 7 Language/Mathematics	Peter Howard
9781862940864	82	Year 8 Language/Mathematics	Peter Howard
9781862941274	117	Maths & Language Problems Level 1 Yrs 1–2	Peter Howard
9781862941281	118	Maths & Language Problems Level 2 Yrs 5–8	Peter Howard
9781862941298	119	Maths & Language Problems Level 1 Yrs 1–2	Peter Howard
9781862941564	146	Challenging Maths Problems Yrs 5–8	
Basic Skills Maths			
9781862940642	65	Basic Skills Maths Level 1 Yrs 1–2	Peter Howard
9781862940659	66	Basic Skills Maths Level 2 Yrs 3–4	Peter Howard
9781862940666	67	Basic Skills Maths Level 3 Yrs 5–8	Peter Howard
9781862940994	95	Practice & Improve Your Maths Yr K–1	Arthur Baillie
9781862941007	96	Practice & Improve Your Maths Yr 2	Arthur Baillie
9781862941014	97	Practice & Improve Your Maths Yr 3	Arthur Baillie
9781862941021	98	Practice & Improve Your Maths Yr 4	Arthur Baillie
9781862941038	99	Practice & Improve Your Maths Yr 5	Arthur Baillie
9781862941045	100	Practice & Improve Your Maths Yr 6	Arthur Baillie
Easy Learn Maths			
9781862942844	145A	Basic Skills Easy– Learn Maths Pre/Kinder A	Valerie Marett
9781862942851	145B	Basic Skills Easy– Learn Maths Pre/Kinder B	Valerie Marett
9781862941410	130	Basic Skills Easy– Learn Maths 1A	Julin Tan
9781862941427	131	Basic Skills Easy– Learn Maths 1B	Julin Tan
9781862941434	132	Basic Skills Easy– Learn Maths 2A	Julin Tan

Coroneos Publications – Titles by Topic

ISBN	No.	Title	Author
9781862941441	133	Basic Skills Easy– Learn Maths 2B	Julin Tan
9781862941458	134	Basic Skills Easy– Learn Maths 3A	Julin Tan
9781862941465	135	Basic Skills Easy– Learn Maths 3B	Julin Tan
9781862941472	136	Basic Skills Easy– Learn Maths 4A	Julin Tan
9781862941489	137	Basic Skills Easy– Learn Maths 4B	Julin Tan
9781862941496	138	Basic Skills Easy– Learn Maths 5A	Julin Tan
9781862941502	139	Basic Skills Easy– Learn Maths 5B	Julin Tan
9781862941519	140	Basic Skills Easy– Learn Maths 6A	Julin Tan
9781862941526	141	Basic Skills Easy– Learn Maths 6B	Julin Tan
9781862941533	142	Basic Skills Easy– Learn Maths 7A Years 6–8	Valerie Marett
9781862941540	143	Basic Skills Easy– Learn Maths 7B Years 6–8	Valerie Marett
Basic Skills Science			
9781862941656	159	Succeeding in Science Yr 1	Graeme Wilson
9781862941663	160	Succeeding in Science Yr 2	Graeme Wilson
9781862941670	161	Succeeding in Science Yr 3	Graeme Wilson
9781862941687	162	Succeeding in Science Yr 4	Graeme Wilson
9781862941694	163	Succeeding in Science Yr 5	Graeme Wilson
9781862941700	164	Succeeding in Science Yr 6	Graeme Wilson
9781862941717	165	Succeeding in Science Yr 7	Graeme Wilson
9781862941809	170	Science and Technology Tests Book 1 Years 3–4	Graeme Wilson
9781862941816	171	Science and Technology Tests Book 2 Years 5–8	Graeme Wilson
Basic Skills Reading Comprehension			
9781862941823	173	Basic Skills – Giant Book of Reading/Comprehension Tests Yrs 5–8	Graeme Wilson
9781862940741	72	Reading/Comprehension Tests for Selective Schools Scholarship Exams Yrs 5–8	Peter Howard
9781862940536	57	Reading/Comprehension Level 1 Yrs 1–2	Peter Howard
9781862940543	58	Reading/Comprehension Level 2 Yrs 3–4	Peter Howard
9781862940550	59	Reading/Comprehension Level 3 Yrs 5–8	Peter Howard
9781862940925	88	First Comprehension Yrs K–3	Peter Howard
9781862941083	103	"Treasure Island" by Robert Louis Stevenson Yrs 3–8	Peter Howard
9781862941090	104	"Black Beauty" by Anna Sewell Yrs 3–8	Peter Howard
9781862941106	105	"Aladdin and "Ali Baba" Yrs K–3	Peter Howard
9781862941182	108	"Wind in the Willows" by Kenneth Grahame Yrs 3–8	Peter Howard
9781862941199	109	Advanced Reading/Comprehension Tests Yrs 7–8	Peter Howard
9781862941267	116	"The Loaded Dog" and other stories by Henry Lawson Yrs 5–8	Peter Howard
9781862941571	147	Advanced Reading/Comprehension Tests Yrs 8–10	Peter Howard
9781862941595	149	"Dot and the Kangaroo" Yrs 3–6	Peter Howard
9781862941625	152	"Aesop's Fables" Years 1–3	Peter Howard
9781862941311	121	"Dad & Dave" by Steele Rudd Yrs 3–8	Peter Howard
Basic Skills Spelling and Vocabulary			
9781862940567	60	Phonic/Reading Spelling Yrs K–3	Peter Howard
9781862940574	61	Spelling/Vocabulary Level 1 Yrs 1–2	Peter Howard
9781862940581	62	Spelling/Vocabulary Level 2 Yrs 3–4	Peter Howard
9781862940604	63	Spelling/Vocabulary Level 3 Yrs 5–8	Peter Howard
9781862941601	150	Vocabulary Builder Level 1 Years 3–4	Peter Howard
9781862941618	151	Vocabulary Builder Level 2 Years 5–8	Peter Howard
Basic Skills English, Writing, Language & Grammar			
9781862940963	92	English Level 1 Yrs 1–2	Peter Howard
9781862940970	93	English Level 2 Yrs 3–4	Peter Howard
9781862940987	94	English Level 3 Yrs 5–8	Peter Howard
9781862940895	85	Grammar & Punctuation Level 1 Yrs 3–4	Peter Howard
9781862940901	86	Grammar & Punctuation Level 2 Yrs 5–8	Peter Howard
9781862941205	110	Advanced Language Tests Yrs 8–10	Peter Howard
9781862941328	122	Advanced English Yrs 8–10	Peter Howard
9781862940710	69	Creative Writing Yrs 3–8	Peter Howard
9781862941366	126	Further creative Writing Yrs 5–8	Peter Howard
9781862941588	148	Advanced Creative Writing Yrs 8–10	Peter Howard
9781862940918	87	Written Expression for Selective Schools/Scholarship Exams Yrs 5–8	Peter Howard
Basic Skills HSIE / SOSE			
9781862940956	91	Activities in Social Studies Yrs 2–8	Peter Howard
9781862941397	129	All About Australia (a New Junior Geography)	Peter Howard
Basic Skills Fun Learning			
9781862941748	156	Fun Learning Activity Book 1 Ages 3–5	Peter Howard
9781862941755	157	Fun Learning Activity Book 2 Ages 6–8	Peter Howard

Coroneos Publications – Titles by Topic

9781862941762	158	Fun Learning Activity Book3 Ages 9–11	Peter Howard
9781862941335	123	Oz Slang All Ages	Peter Howard
9781862942141	204	More Oz Slang	Peter Howard

Basic Skills Early Years

9781862941342	124	Introductory Comprehension Yrs K–3	Peter Howard
9781862941373	127	Pre-School Activities 1	Peter Howard
9781862941380	128	Pre-School Activities 2	Peter Howard
9781862940703	68	Reading Practice Yrs 1–3	Peter Howard
9781862940772	75	My First Reading 1000 Reading/Spelling Words yrs K–3	Peter Howard
9781862940819	77	Fun with Words Book 1 (colour) Yrs K–3	Peter Howard
9781862940826	78	Fun with Words Book 2 (colour) Yrs K–3	Peter Howard
9781862940833	79	Counting Practice Yrs K and Under	Peter Howard
9781862940840	80	First Reading Yrs k and Under	Peter Howard
9781862940932	89	First Creative Writing Yrs K–3	Peter Howard
9781862940949	90	Picture words Yrs K and under	Peter Howard
9781862941212	111	First Phonics Yrs K–3	Peter Howard
9781862941229	113	First phonic reading Yrs K–3	Peter Howard
9781862941236	114	Introductory Creative Writing Yrs K–3	Peter Howard
9781862941250	115	First Words Yrs K–3	Peter Howard
9781862941304	120	Second words Yrs K–3	Peter Howard
9781862940871	83	First handwriting Yrs K–2	Peter Howard
9781862940888	84	Handwriting Practice yrs 3–4	Peter Howard

Early Basic Skills NSW font

9781921565564	266	Early Basic Skills Single Sounds – using NSW font	D.J. Ferguson
9781921565571	267	Early Basic Skills Simple Words & Sentences – using NSW font	D.J. Ferguson
9781921565588	268	Early Basic Skills Blending Consonants – using NSW font	D.J. Ferguson
9781921565595	269	Early Basic Skills Using Diagraphs – using NSW font	D.J. Ferguson
9781921565601	270	Early Basic Skills Written Text – Punctuation & Grammar – using NSW font	D.J. Ferguson

Early Basic Skills Victorian font

9781921565618	366	Early Basic Skills Single Sounds – using Victorian font	D.J. Ferguson
9781921565625	367	Early Basic Skills Simple Words & Sentences – using Victorian font	D.J. Ferguson
9781921565632	368	Early Basic Skills Blending Consonants – using Victorian font	D.J. Ferguson
9781921565649	369	Early Basic Skills Using Diagraphs – using Victorian font	D.J. Ferguson
9781921565656	370	Early Basic Skills Written Text – Punctuation & Grammar – using Victorian font	D.J. Ferguson

Early Learning Skills (old editions)

9781862941779	166	Early Learning Skills Book 1 Pre School/Kinder - Single Sounds	Denise Tyras
9781862941786	167	Early Learning Skills Book 2 Years K–1 - Three Letter Words	Denise Tyras
9781862941793	168	Early Learning Skills Book 3 Years 1–2 - Blends	Denise Tyras
9781862941854	172	Grammar, Punctuation and Vocabulary Years 1/2	Denise Tyras

DON ROBENS TITLES

Spelling

9781862941946	188	Spelling Year 1	Don Robens
9781862941953	189	Spelling Year 2	Don Robens
9781862941960	190	Spelling Year 3	Don Robens
9781862941977	191	Spelling Year 4	Don Robens
9781862941984	192	Spelling Year 5	Don Robens
9781862941991	193	Spelling Year 6	Don Robens

Creative Writing

9781862942028	196	Writing in Text Types Year 3	Don Robens
9781862942035	197	Writing in Text Types Year 4	Don Robens
9781862942042	198	Writing in Text Types Year 5	Don Robens
9781862942059	199	Writing in Text Types Year 6	Don Robens

Comprehension

9781921565755	248	Comprehension Year 3	Don Robens
9781921565762	249	Comprehension Year 4	Don Robens
9781921565779	250	Comprehension Year 5	Don Robens
9781921565786	251	Comprehension Year 6	Don Robens

Coroneos Publications – Titles by Topic

AUSTRALIAN HOMESCHOOLING

Mathematics			
9781862942158	501	Learning Multiplication 1 – Australian Homeschooling	Carmel Musumeci
9781862942165	502	Learning Multiplication 2 – Australian Homeschooling	Carmel Musumeci
9781921565298	531	Practise Your Addition and Subtraction to 20	Marett & Musumeci
Social Studies			
9781862942509	503	Succeeding in Social Studies K	Valerie Marett
9781862942516	504	Succeeding in Social Studies 1	Valerie Marett
9781862942523	505	Succeeding in Social Studies 2	Valerie Marett
9781862942530	506	Succeeding in Social Studies 3	Valerie Marett
9781862942547	507	Succeeding in Social Studies 4	Valerie Marett
9781862942554	508	Succeeding in Social Studies 5	Valerie Marett
9781862942561	509	Succeeding in Social Studies 6	Valerie Marett
9781921565212	523	Australian History 1901–1945	Valerie Marett
9781921565335	535	Australian Government	Valerie Marett
English			
9781862942219	510	Successful English 1	Valerie Marett
9781862942226	511	Successful English 2	Valerie Marett
9781862942479	512	Successful English 3A	Valerie Marett
9781862942486	513	Successful English 3B	Valerie Marett
9781921565250	527	Successful English 4A	Valerie Marett
9781921565267	528	Successful English 4B	Valerie Marett
9781921565342	536	Successful English 5A	Valerie Marett
9781921565359	537	Successful English 5B	Valerie Marett
Phonics			
9781921565229	524	Revise Your Phonics 1	Valerie Marett
9781921565236	525	Revise Your Phonics 2	Valerie Marett
Spelling			
9781862942233	514	Successful Spelling K	Valerie Marett
9781862942240	515	Successful Spelling 1	Valerie Marett
9781862942257	516	Successful Spelling 2	Valerie Marett
9781862942493	517	Successful Spelling 3	Valerie Marett
9781921565304	532	Successful Spelling 4	Valerie Marett
9781921565663	538	Successful Spelling 5	Valerie Marett
9781862942837	177	Learn to Read, Write & Spell 1 – Yrs K–1	Valerie Marett
9781862942820	178	Learn to Read, Write & Spell 2 – Yrs K–1	Valerie Marett
9781862942769	179	Learn to Read, Write & Spell 3 – Yrs 1–3	Valerie Marett
9781862942776	180	Learn to Read, Write & Spell 4 – Yrs 1–3	Valerie Marett
9781862942783	181	Learn to Read, Write & Spell 5 – Yrs 1–4	Valerie Marett
9781862942790	182	Learn to Read, Write & Spell 6 – Yrs 3–8	Valerie Marett
9781862942806	183	Learn to Read, Write & Spell 7 – Yrs 5–10	Valerie Marett
Testing			
9781862942172	518	Test your Maths K–1	Valerie Marett
9781862942189	519	Test Your Maths 2	Valerie Marett
9781921565243	526	Test Your Maths 3	Valerie Marett
9781921565328	534	Test Your Maths 4	Valerie Marett
9781921565687	540	Test Your Maths 5	Valerie Marett
9781862942196	520	Test your Spelling & English K–1	Valerie Marett
9781862942202	521	Test your Spelling & English 2	Valerie Marett
9781862942578	522	Test your Reading & Phonics K–3	Valerie Marett
9781921565274	529	Test Your Spelling Years 3 & 4	Valerie Marett
9781921565281	530	Test Your English 3	Valerie Marett
9781921565311	533	Test Your English 4	Valerie Marett
9781921565670	539	Test Your English 5	Valerie Marett
Science			
9781921565694	541	Secondary Science 7A	Frank Marett
9781921565700	542	Secondary Science 7B	Frank Marett
9781921565717	543	Secondary Science 7C	Frank Marett
9781921565724	544	Secondary Science 7D	Frank Marett
9781921565731	545	Secondary Science 8A	Frank Marett
9781921565748	546	Secondary Science 8B	Frank Marett

Coroneos Publications – Titles by Topic

MATHEMATICS

Secondary Texts

9781862940017	2	Year 9,10 Preparing for Year 11 Mathematics	Jim Coroneos
9781862940062	7	Year 11, 3 Unit Mathematics Course	Jim Coroneos
9781862940079	8	Year 12 3 Unit maths Course	Jim Coroneos
9781862940154	16	Complete 2 Unit Mathematics Course in 1 Volume	Jim Coroneos
9781862940178	18	Simplified 3 Unit Maths Course Vol 2	Jim Coroneos
9781862940185	19	Revised 4 unit Maths Course	Jim Coroneos
9781862940192	20	Supplement for 4 Unit Maths (large page)	Jim Coroneos

Past Papers and Solutions

9781921565380	35	HSC Mathematics Past Papers1990–2009 with Worked Solutions	Jim Coroneos & Others
9781921565373	36	HSC Extension 1 Mathematics Past Papers 1990–2009 with Worked Solutions	Jim Coroneos & Others
9781921565366	37	HSC Extension 2 Mathematics Past Papers 1990–2009 with Worked Solutions	Jim Coroneos & Others
9781921565397	175	HSC General Mathematics Past Papers 2001–2009 with Worked Solutions	Jim Coroneos & Others
9781921565427	206	HSC Mathematics 2001–2009 Past Papers & Worked Solutions	Jim Coroneos & Others
9781921565410	207	HSC Mathematics Extension 1 2001–2009 Past Papers & Worked Solutions	Jim Coroneos & Others
9781921565403	208	HSC Mathematics Extension 2 2001–2009 Past Papers & Worked Solutions	Jim Coroneos & Others

Past Papers and Solutions (Older Series)

9781862940321	33	Past HSC 2 Unit/3 Unit Maths Papers from 1967 with Answers	Jim Coroneos
9781862940338	34	Past HSC 3 Unit/4 Unit Maths Papers from 1967 with Answers	Jim Coroneos
9781862940673	47	Past NSW SC Maths Papers Advanced Level from 1983 with Answers/Solutions	Jim Coroneos
9781862940680	48	Past NSW SC Maths Papers Intermediate Level from 1983 with Answers/Solutions	Jim Coroneos
9781862941113	106	Past NSW SC Maths Paper General Level from 1988 with Answers/Solutions	Jim Coroneos

Practice (Specimen) Papers

9781862940260	27	Mathematics Practice papers with Answers	Jim Coroneos
9781862940277	28	Mathematics Extension 1 Practice Papers with Answers	Jim Coroneos
9781862940284	29	Mathematics Practice papers with Worked solutions	Jim Coroneos
9781862940291	30	Mathematics Extension 1 Practice Papers with Worked Solutions	Jim Coroneos
9781862940307	31	Mathematics Extension 1 Practice papers Plus Worked Solutions	Jim Coroneos
9781862940376	38	Short Question Ordinary/Credit Level Maths Papers with Answers	Jim Coroneos
9781862940383	39	Solutions to Short Question Ordinary Credit Level Maths Papers	Jim Coroneos
9781862940406	41	Short Question Advanced Level Maths Papers with Worked Solutions	Jim Coroneos
9781862940420	43	New Course Advanced Level Maths Papers with Answers	Jim Coroneos
9781862940437	44	New Course intermediate Level Maths Papers with Answers	Jim Coroneos
9781862940444	45	New Course Advanced Level Maths Papers with Worked Solutions	Jim Coroneos
9781862940451	46	New Course Intermediate Level Maths Papers with Worked Solutions	Jim Coroneos

Study Skills

9781862940000	1	How to Study Effectively Yrs 4–8	Coroneos & Smith

Coroneos Publications – Titles by Topic

EXCELLENCE SERIES

English

ISBN	No.	Title	Author
9781862942356	211	Excellence in English Year 1	Peter Howard
9781862942363	212	Excellence in English Year 2	Peter Howard
9781862942370	213	Excellence in English Year 3	Peter Howard
9781862942387	214	Excellence in English Year 4	Peter Howard
9781862942394	215	Excellence in English Year 5	Peter Howard
9781862942400	216	Excellence in English Year 6	Peter Howard
9781875695256		Excellence in English for Secondary Students	Peter Howard
9781875695331		English Skills in Use for Secondary Students	Peter Howard

Reading Skills

ISBN	No.	Title	Author
9781862942417	217	Excellence in Reading Skills Year 1	Peter Howard
9781862942424	218	Excellence in Reading Skills Year 2	Peter Howard
9781862942431	219	Excellence in Reading Skills Year 3	Peter Howard
9781862942448	220	Excellence in Reading Skills Year 4	Peter Howard
9781862942455	221	Excellence in Reading Skills Year 5	Peter Howard
9781862942462	222	Excellence in Reading Skills Year 6	Peter Howard
9781875695485		Practical Reading and Writing Skills	Peter Howard
9781875695324		Reading and Writing Skills for secondary students	Peter Howard
9781875695904		Excellence in Written Expression for secondary students	Peter Howard

Literacy

ISBN	No.	Title	Author
9781921565007	223	Excellence in Literacy Year 1	Peter Howard
9781921565014	224	Excellence in Literacy Year 2	Peter Howard
9781921565021	225	Excellence in Literacy Year 3	Peter Howard
9781921565038	226	Excellence in Literacy Year 4	Peter Howard
9781921565045	227	Excellence in Literacy Year 5	Peter Howard
9781921565052	228	Excellence in Literacy Year 6	Peter Howard
9781875695532		Practical Word Skills for Secondary Students	Peter Howard
9781875695652		Practical Spelling Skills	Judith Hall & Ron King
9781862942103		Middle School Vocabulary	Peter Howard
9781862942110		Dictionary Skills for Year 3/4	Peter Howard

Mathematics

ISBN	No.	Title	Author
9781921565069	229	Excellence in Maths Year 1	Peter Howard
9781921565076	230	Excellence in Maths Year 2	Peter Howard
9781921565083	231	Excellence in Maths Year 3	Peter Howard
9781921565090	232	Excellence in Maths Year 4	Peter Howard
9781921565106	233	Excellence in Maths Year 5	Peter Howard
9781921565113	234	Excellence in Maths Year 6	Peter Howard
9781875695386		Excellence in Maths 1 for secondary students	Fisher & Howard
9781875695416		Excellence in Maths 2 for secondary students	George Fisher

General Ability

ISBN	No.	Title	Author
9781862942066	209	Excellence in General Ability Year 4	Peter Howard
9781862942073	210	Excellence in General Ability Year 5	Peter Howard

PHONICS SERIES

ISBN	No.	Title	Author
9781862942080	261	Step by Step Phonics 1 NSW version ***NSW Foundation Handwriting***	Peter Howard
9781862942097	262	Step by Step Phonics 2 NSW version ***NSW Foundation Handwriting***	Peter Howard
9781921565199	263	Step by Step Phonic Reading 1 NSW version ***NSW Foundation Handwriting***	Peter Howard
9781921565205	264	Step by Step Phonic Reading 2 NSW version ***NSW Foundation Handwriting***	Peter Howard
9781862942127	361	Step by Step Phonics 1 Vic version *Victorian Modern Cursive*	Peter Howard
9781862942134	362	Step by Step Phonics 2 Vic version *Victorian Modern Cursive*	Peter Howard
9781862942332	363	Step by Step Phonic Reading 1 Vic version *Victorian Modern Cursive*	Peter Howard
9781862942349	364	Step by Step Phonic Reading 2 Vic version *Victorian Modern Cursive*	Peter Howard